AF368954

# Energy Harvesters and Self-powered Sensors for Smart Electronics

# Energy Harvesters and Self-powered Sensors for Smart Electronics

Editors

**Qiongfeng Shi**
**Huicong Liu**

MDPI • Basel • Beijing • Wuhan • Barcelona • Belgrade • Manchester • Tokyo • Cluj • Tianjin

*Editors*
Qiongfeng Shi
Department of Electrical and
Computer Engineering
National University of Singapore
Singapore

Huicong Liu
School of Mechanical and
Electrical Engineering
Soochow University
Suzhou
China

*Editorial Office*
MDPI
St. Alban-Anlage 66
4052 Basel, Switzerland

This is a reprint of articles from the Special Issue published online in the open access journal *Micromachines* (ISSN 2072-666X) (available at: www.mdpi.com/journal/micromachines/special_issues/energy_harvesters_self-powered_sensors).

For citation purposes, cite each article independently as indicated on the article page online and as indicated below:

LastName, A.A.; LastName, B.B.; LastName, C.C. Article Title. *Journal Name* **Year**, *Volume Number*, Page Range.

ISBN 978-3-0365-2675-1 (Hbk)
ISBN 978-3-0365-2674-4 (PDF)

# Contents

# About the Editors

**Qiongfeng Shi**

Qiongfeng Shi has been a Research Fellow in the Department of Electrical and Computer Engineering at the National University of Singapore (NUS) since 2018. He received his B.Eng. degree from the Department of Electronic Engineering and Information Science at the University of Science and Technology of China (USTC) in 2012, and received his Ph.D. degree from the Department of Electrical and Computer Engineering at the National University of Singapore in 2018. His research interests include energy harvesters, nanogenerators, sensors, MEMS, wearable electronics, human-machine interfaces, and intelligent systems, etc. Dr. Shi has authored and co-authored more than 80 international journal papers and conference papers in his research fields, which have received more than 3200 citations with an H-index of 35 according to Google Scholar. Until now, Dr. Shi has served as a Guest Editor for several MDPI journals, and as a reviewer for multiple international journals.

**Huicong Liu**

Huicong Liu is currently a Professor with the Robotics and Microsystems Center, School of Mechanical and Electrical Engineering, Soochow University, China. She received the B.Eng. and M.Sc. Degrees from the Department of Mechanical Engineering, University of Science and Technology Beijing, China, in 2006 and 2008, respectively, and the Ph.D. degree from the Department of Mechanical Engineering, National University of Singapore (NUS), in 2013. She was a Research Fellow with the Department of Electrical and Computer Engineering, NUS, from 2012 to 2013. Her research interests are vibration-based MEMS/NEMS energy harvesters, self-powered MEMS/NEMS system for IoTs, flexible and intelligent sensing for human-machine interface. Prof. Liu has contributed 2 academic books, and more than 100 academic papers. Her work has been cited for more than 2800 times with H-index of 29. She was the co-chairman of the 2nd International Conference on Vibration Energy Harvesting (VEH 2019), TPC member of IEEE NEMS 2018-2019, 2021, and TPC member of Transducers 2021.

# Preface to "Energy Harvesters and Self-powered Sensors for Smart Electronics"

Recently, we have witnessed the revolutionary innovation and flourishing advancement of the Internet of Things (IoT), which will maintain a strong momentum even more with the gradual rollout of the fifth generation (5G) wireless network and the rapid development of personal healthcare electronics. Enabled by the ultrahigh data communication rate of 5G technology, various IoT systems can be envisioned by linking numerous multifunctional electronic devices together in an integrated and interconnected network, such as in smart homes, smart buildings, industry 4.0, smart cities, unmanned shops, wearable body area networks, point-of-care diagnosis and treatment, etc. In these complicated systems with widely distributed device nodes, the energy supply in the IoT era is gradually migrating from the regularly replaced battery-based supply mode toward the sustainable in situ supply mode. Compared to batteries, energy harvesting technologies that scavenge available energies from the ambient surroundings exhibit great merits in supplying energy to IoT systems, e.g., highly extended and unlimited lifetime, great portability, flexible/stretchable compatibility, and sustainability. Therefore, various energy harvesting technologies and devices are undergoing rapid and significant innovation, providing key functionalities in diverse IoT systems as energy harvesters and/or self-powered sensors. Accordingly, this book showcases one editorial and nine research articles that focus on the novel development of energy harvesters/self-powered sensors and their broad applications in IoT systems and smart electronics. The research articles explore the following aspects of energy harvesting technologies, such as electromagnetic energy harvesters, piezoelectric energy harvesters, and hybrid energy harvesters: mechanism design, structural optimization, performance improvement, and a wide range of energy harvesting and self-powered monitoring applications.

We would like to express our sincere appreciation to all the authors for contributing their papers to this Special Issue Book. Meanwhile, we would also like to thank all the reviewers for dedicating their valuable time and helping to ensure the quality of this Special Issue Book.

**Qiongfeng Shi, Huicong Liu**
*Editors*

*micromachines*

*Editorial*

# Special Issue on Energy Harvesters and Self-Powered Sensors for Smart Electronics

Qiongfeng Shi [1,*] and Huicong Liu [2,*]

1   Department of Electrical and Computer Engineering, National University of Singapore, 4 Engineering Drive 3, Singapore 117583, Singapore
2   School of Mechanical and Electric Engineering, Jiangsu Provincial Key Laboratory of Advanced Robotics, Soochow University, Suzhou 215021, China
*   Correspondence: qiongfeng@u.nus.edu (Q.S.); hcliu078@suda.edu.cn (H.L.)

**Citation:** Shi, Q.; Liu, H. Special Issue on Energy Harvesters and Self-Powered Sensors for Smart Electronics. *Micromachines* **2021**, 12, 1455. https://doi.org/10.3390/mi12121455

Received: 22 November 2021
Accepted: 25 November 2021
Published: 26 November 2021

**Publisher's Note:** MDPI stays neutral with regard to jurisdictional claims in published maps and institutional affiliations.

In recent years, we have witnessed the revolutionary innovation and flourishing advancement of the Internet of things (IoT), which will maintain a strong momentum even more with the gradual rollout of the fifth generation (5G) wireless network and the rapid development of personal healthcare electronics. Enabled by the ultrahigh data communication rate of 5G technology, various IoT systems can be envisioned by linking numerous multifunctional electronic devices together in an integrated and interconnected network, such as in smart homes, smart buildings, industry 4.0, smart cities, unmanned shops, wearable body area networks, point-of-care diagnosis and treatment, etc. In these complicated systems with widely distributed device nodes, the energy supply in the IoT era is gradually migrating from the regularly replaced battery-based supply mode toward the sustainable in situ supply mode. Compared to batteries, energy harvesting technologies that scavenge available energies from the ambient surroundings exhibit great merits in supplying energy to IoT systems, e.g., highly extended and unlimited lifetime, great portability, flexible/stretchable compatibility, and sustainability. Therefore, various energy harvesting technologies and devices are undergoing rapid and significant innovation, providing key functionalities in diverse IoT systems as energy harvesters and/or self-powered sensors. Accordingly, this Special Issue showcases nine research articles that focus on the novel development of energy harvesters/self-powered sensors and their broad applications in IoT systems and smart electronics. The research articles in this Special Issue explore the following aspects of energy harvesting technologies such as electromagnetic energy harvesters, piezoelectric energy harvesters, and hybrid energy harvesters: mechanism design, structural optimization, performance improvement, and a wide range of energy harvesting and self-powered monitoring applications.

In terms of the electromagnetic energy harvesters, Li et al. [1] designed a magnetically coupled electromagnetic energy harvester composed of two spring-connected magnets, one free magnet and coils, for scavenging the low-frequency human motion energy. The main advantage of the magnetic-spring structure is the weakened repulsive force, enabling effective low-frequency energy harvesting. Under handshaking (0.2 g and 3.4 Hz) and leg movement excitation, the energy harvester obtained a maximum output voltage of 0.6 V and a maximum output power of 26 mW. Cheng et al. [2] investigated in detail the characteristics of a diamagnetically stabilized levitation structure by both simulation and experiment. They found out that the symmetric monostable levitation system is more stable, and further explored the optimal parameters, achieving an output voltage of 250.69 mV and output power of 86.8 μW. Huang et al. [3] also proposed a magnetic-coupled electromagnetic energy harvester with nonlinear features and a broad bandwidth through the strong coupling between an iron core and vibrating magnets. After optimization, the energy harvester realized a broad operation bandwidth of 17–30 Hz and a high output power of 174 mW (at 35 Ω) that was able to light up 360 parallel-connected LEDs.

As for the piezoelectric energy harvesters, Song et al. [4] studied the enhancement effect of two piezoelectric energy harvesters in the context of water flow-induced vibration. The performance correlation of the upstream and the downstream piezoelectric energy harvester was theoretically and experimentally investigated. Thereupon, optimization was performed to boost the total output power which was significantly higher than that of a single piezoelectric energy harvester. Huang et al. [5] proposed an automatic data-driven strategy to optimize the design of piezoelectric energy harvesters, based on the algorithm of generalized pattern search (GPS), which can effectively save designers' effort and achieve more precise parameters. Employing the optimal length and thickness, the derived energy harvester showed an increment of 371% and 1000% in output power and normalized power density, respectively. Han et al. [6] developed a bimetallic-cantilever-triggered piezoelectric energy harvester for self-powered and automatic heat-event monitoring. If the temperature reaches a preset threshold, the bimetallic cantilever will detach from the piezoelectric energy harvester, causing the resonance and output generation of the piezoelectric energy harvester. Under continuous heating, such temperature-triggered cycles were repeated with electric power generation. Zhang et al. [7] investigated a rope-driven piezoelectric energy harvester for broadband and low-frequency energy harvesting, which consists of a low-frequency cantilever and a high-frequency cantilever connected by a rope. After systematic simulation and experiment, the characteristics of the rope-driven energy harvester were validated, indicating that its performance could be conveniently adjusted by the rope margin or stiffness. Huang et al. [8] presented a multi-DOF piezoelectric energy harvester with a wide bandwidth using the frequency interval shortening design. Configured with one straight cantilever and two U-shaped cantilevers with proof masses, the energy harvester exhibited five low resonances, i.e., 13, 15, 18, 21 and 24 Hz, constituting a broad operation bandwidth at low frequencies.

Based on the hybridized mechanism of electromagnetics and piezoelectricity, Li et al. [9] proposed an electromagnetic hybridized scheme to improve the output performance of a piezoelectric energy harvester. The mass block of a piezoelectric cantilever was replaced by an alternating magnet array and a coil array was added beside the magnet array, forming the hybrid energy harvester. It achieved a power density of 3.53 mW/cm$^3$ at the acceleration of 0.3 g and frequency of 18.6 Hz, which is 686 times higher than that of a basic cantilever piezoelectric energy harvester. In addition, the hybrid energy harvester also exhibited excellent performance in charging capacitors or batteries, e.g., a 2.2 mF capacitor was charged up to 8 V within 17 s, showing great promise in developing self-powered electronics in IoT and wearable applications.

We would like to express our sincere appreciation to all the authors for submitting their papers to this Special Issue. Meanwhile, we would also like to thank all the reviewers for dedicating their valuable time and helping to ensure the quality of this Special Issue.

**Funding:** This work is funded by the National Science Foundation of China (Grant No. 51875377).

**Conflicts of Interest:** The authors declare no conflict of interest.

## References


1.  Li, X.; Meng, J.; Yang, C.; Zhang, H.; Zhang, L.; Song, R. A Magnetically Coupled Electromagnetic Energy Harvester with Low Operating Frequency for Human Body Kinetic Energy. *Micromachines* **2021**, *12*, 1300. [CrossRef]
2.  Cheng, S.; Li, X.; Wang, Y.; Su, Y. Levitation Characteristics Analysis of a Diamagnetically Stabilized Levitation Structure. *Micromachines* **2021**, *12*, 982. [CrossRef] [PubMed]
3.  Huang, M.; Li, Y.; Feng, X.; Tang, T.; Liu, H.; Chen, T.; Sun, L. A Magnetic-Coupled Nonlinear Electromagnetic Generator with Both Wideband and High-Power Performance. *Micromachines* **2021**, *12*, 912. [CrossRef] [PubMed]
4.  Song, R.; Hou, C.; Yang, C.; Yang, X.; Guo, Q.; Shan, X. Modeling, Validation, and Performance of Two Tandem Cylinder Piezoelectric Energy Harvesters in Water Flow. *Micromachines* **2021**, *12*, 872. [CrossRef] [PubMed]
5.  Huang, Y.; Yi, Z.; Hu, G.; Yang, B. Data-Driven Optimization of Piezoelectric Energy Harvesters via Pattern Search Algorithm. *Micromachines* **2021**, *12*, 561. [CrossRef] [PubMed]
6.  Han, R.; Wang, N.; He, Q.; Wang, J.; Li, X. An Energy Harvester with Temperature Threshold Triggered Cycling Generation for Thermal Event Autonomous Monitoring. *Micromachines* **2021**, *12*, 425. [CrossRef]

7. Zhang, J.; Lin, M.; Zhou, W.; Luo, T.; Qin, L. Modeling of a Rope-Driven Piezoelectric Vibration Energy Harvester for Low-Frequency and Wideband Energy Harvesting. *Micromachines* **2021**, *12*, 305. [CrossRef] [PubMed]
8. Huang, X.; Zhang, C.; Dai, K. A Multi-Mode Broadband Vibration Energy Harvester Composed of Symmetrically Distributed U-Shaped Cantilever Beams. *Micromachines* **2021**, *12*, 203. [CrossRef] [PubMed]
9. Li, Z.; Xin, C.; Peng, Y.; Wang, M.; Luo, J.; Xie, S.; Pu, H. Power Density Improvement of Piezoelectric Energy Harvesters via a Novel Hybridization Scheme with Electromagnetic Transduction. *Micromachines* **2021**, *12*, 803. [CrossRef] [PubMed]

*Article*

# A Magnetically Coupled Electromagnetic Energy Harvester with Low Operating Frequency for Human Body Kinetic Energy

Xiang Li [1], Jinpeng Meng [1], Chongqiu Yang [1], Huirong Zhang [1], Leian Zhang [1] and Rujun Song [1,2,*]

1. School of Mechanical Engineering, Shandong University of Technology, Zibo 255049, China; lixiangsdut@163.com (X.L.); mengjinpengvip@126.com (J.M.); yangcq@sdut.edu.cn (C.Y.); fdzhanghr@163.com (H.Z.); ziaver@163.com (L.Z.)
2. Shenzhen Research Institute of City University of Hong Kong, Shenzhen 518057, China
* Correspondence: songrujun@sdut.edu.cn

**Abstract:** In this paper, a magnetically coupled electromagnetic energy harvester (MCEEH) is proposed for harvesting human body kinetic energy. The proposed MCEEH mainly consists of a pair of spring-connected magnets, coils, and a free-moving magnet. Specifically, the interaction force between the magnets is repulsive. The main feature of this structure is the use of a magnetic-spring structure to weaken the hardening response caused by the repulsive force. The magnetic coupling method enables the energy harvester system to harvest energy efficiently at low frequency. The MCEEH is experimentally investigated for improving energy harvesting efficiency. Under harmonic excitation with an acceleration of 0.5 g, the MCEEH reaches resonance frequency at 8.8 Hz and the maximum output power of the three coils are 5.2 mW, 2.8 mW, and 2.5 mW, respectively. In the case of hand-shaking excitation, the generator can obtain the maximum voltage of 0.6 V under the excitation acceleration of 0.2 g and the excitation frequency of 3.4 Hz. Additionally, a maximum instantaneous power can be obtained of about 26 mW from the human body's kinetic energy.

**Keywords:** vibration energy harvesting; electromagnetic energy harvester; magnetic coupling; human body kinetic energy

**Citation:** Li, X.; Meng, J.; Yang, C.; Zhang, H.; Zhang, L.; Song, R. A Magnetically Coupled Electromagnetic Energy Harvester with Low Operating Frequency for Human Body Kinetic Energy. *Micromachines* **2021**, *12*, 1300. https://doi.org/10.3390/mi12111300

Academic Editors: Qiongfeng Shi and Huicong Liu

Received: 26 September 2021
Accepted: 20 October 2021
Published: 22 October 2021

**Publisher's Note:** MDPI stays neutral with regard to jurisdictional claims in published maps and institutional affiliations.

## 1. Introduction

In recent years, the disadvantages of chemical batteries as power supply are obvious. These kinds of shortages including low-level capacity and lifespan, heavy metal pollution, etc. Nowadays, there is a new field that has gained broad concern, and that is the conversion of environmental energy into electric energy [1,2]. These energy conversion technologies can avoid the aforementioned disadvantages of chemical battery power and reduce the dependence of small electronic devices on batteries [3–5]. According to the conductive mechanism, the conversion can be divided into piezoelectric [6–9], electromagnetic [10–14], and electrostatic [15–18]. The electromagnetic vibration energy harvester has simple structure, small internal resistance, high current, and easy processing compared to other types of energy harvesters, thus receiving more and more research and attention [19–22]. Conventional electromagnetic generators generally include stators, rotors, bearings, etc. Their complex structure is not conducive to miniaturization [23]. Therefore, researchers have developed a new type of electromagnetic energy harvester. For example, Saha et al. [24] designed an electromagnetic energy harvester (EMEH), which was composed of magnets fixed at the ends of the cylinder, a magnet suspended in the middle of the cylinder by the repulsive force, and a coil wound on the outer surface of the cylinder. The middle magnet can move freely. When the cylinder is subjected to external vibration excitation, the central magnet will vibrate up and down, and the current will generate in the coil. Mann et al. [25,26] designed a novel energy harvesting device that used magnetic restoring forces to harvest energy from the nonlinear oscillations. In addition, Mann innovated and created a nonlinear energy harvester with a bistable potential well. Lee et al. [27]

designed an electromagnetic energy harvester that could generate higher output power at a lower frequency. Foisal et al. [28] proposed two kinds of electromagnetic multi-frequency converter array models and used the magnetic spring technique to harvest energy from low-frequency vibrations. Fan et al. [29,30] presented a monostable electromagnetic energy harvester. The energy harvester has a typical softening response and can move the working frequency band to the left. It is very suitable for extracting energy from ultra-low frequency excitation. Further studies by Fan also presented a two-degree-of-freedom (2-DOF) electromagnetic energy harvester (EMEH). Both experiments and simulation verification results showed that the energy harvester can increase the output power and expand the working bandwidth. Zhu et al. [31] investigated a broadband compact electromagnetic energy harvester with a coupled bistable structure. Compared with conventional bistable and linear energy harvesters, this harvester can generate a larger power output. Masana et al. [32] investigated the relative performance of the energy harvester under different monostable and bistable conditions. Results show that the bistable harvester with shallow potential well produced large power in the low-frequency range. Munaz et al. [33] developed an electromagnetic energy harvester using a multi-pole magnet to obtain high power output within a limited volume. Halim et al. [34] designed a frequency conversion harvester that used the collision of small balls to convert low-frequency vibration to a high-frequency vibration. However, energy loss existed in the collision process.

Accordingly, the above harvester has not been studied together with the magnetic levitation structure under magnetic spring structure. Consequently, we propose a magnetically coupled electromagnetic energy harvester (MCEEH) to harvest vibration energy at low-frequency. The proposed MCEEH mainly consists of a pair of spring-connected magnets, coils, and a free-moving magnet. Specifically, the interaction force between the magnets is repulsive. The main feature of this structure is the use of a magnetic-spring structure to weaken the hardening response caused by the repulsive force. This MCEEH has the advantages of adjustable working frequency and high low-frequency energy harvest efficiency. Moreover, this paper also studies the effects of excitation frequency, excitation acceleration, and spring stiffness on the efficiency of energy harvesters under harmonic excitation. The human body kinetic energy experiment was conducted to expand the application of electromagnetic energy harvesters, including: hand-shaking excitation experiment, leg excitation experiment, and backpack experiment. The experimental results show that the MCEEH can be used to power various microelectronic products, e.g., wrist watches, sensors, hearing aids, etc.

## 2. Physical Model and Experimental Method

In this paper, a magnetically coupled electromagnetic energy harvester is proposed and investigated, as shown in Figure 1. The energy harvester was composed of a cylinder, magnet, spring, sliding table, plug, and aluminum rod. The plug was fixed at the end of the cylinder, and two aluminum rods were fixed at the center of the plug. Furthermore, one end of the spring was fixed to the plug and the other end was fixed to the slide. The sliding table was hollowed and slid freely across the aluminum rod. The aluminum rod connected the whole device internally. Three groups of magnets were placed on the sliding table, two of which were placed on the sliding table at each end and the other central magnet was levitated by the repulsive force of the two magnets from both sides. Three groups of coils were wrapped around the outer surface of the cylindrical cavity.

Based on the physical model in Figure 1, the experimental prototype of the MCEEH and test platform was fabricated and assembled respectively, as showed in Figure 2. In detail, the length, outer diameter, and wall thickness of the cylinder were 120 mm, 24 mm, and 2 mm, respectively. Three sets of 800 coils were wrapped around the outside of the cylinder, and the diameter of the coil was 0.12 mm. The internal resistance of the coil and resistance of the external resistance had the same resistance value of 220 $\Omega$. The magnet was round with a hole of about 5 mm. The diameter and thickness of the magnet were 20 mm and 5 mm, respectively. The residual magnetic induction density was 1.32 T. The masses of the upper, middle, and lower magnets were 30 g, 10 g, and 30 g, respectively. The elastic stiffness of the spring was 29.16 g/mm. The experimental system consisted of a vibration exciter, power amplifier, vibration controller, data acquisition card, acceleration sensor, and computer. The acceleration sensor was mounted on the aluminum base plate with a permanent magnet. The data acquisition card was connected to the computer to collect the experimental data such as the real-time voltage at both ends of the load resistor of the MCEEH device. It is worth noting that the cylinder was vertically installed on the bottom plate in the experiment. The output performance of the energy harvester device was investigated by changing the excitation frequency, excitation acceleration, and spring stiffness, respectively.

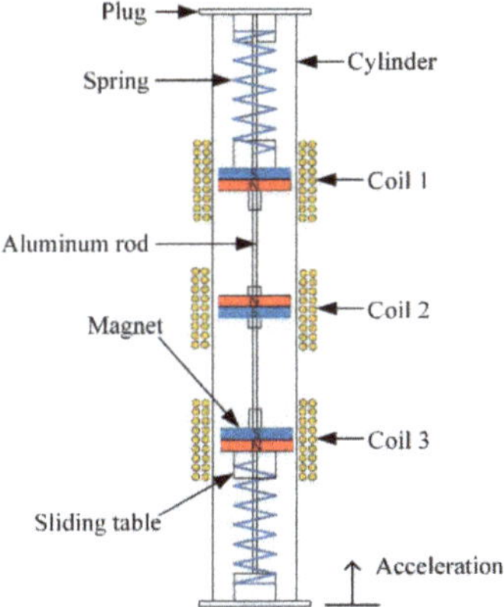


**Figure 1.** Schematic diagram of the magnetically coupled electromagnetic energy harvester.

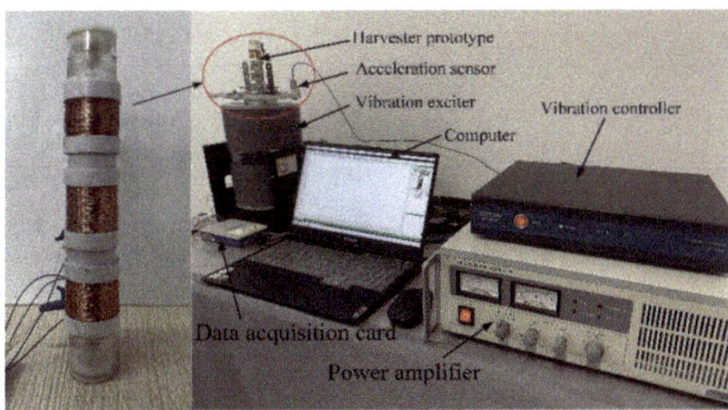


**Figure 2.** Experimental platform and data acquisition system.

## 3. Results and Discussion

### 3.1. Electrical Output under Harmonic Excitations

This section describes the output characteristics of the MCEEH under vibration excitation. Experimental investigation of the output characteristics of the MCEEH used the test platform in Figure 2. First, the output voltage and output power of the MCEEH were tested with varied excitation frequencies. The excitation acceleration was 0.5 g. Figure 3a is the curve of the output voltage of the MCEEH with the excitation frequency, and Figure 3b is the curve of the output power of the MCEEH with the excitation frequency. It is shown in Figure 3 that the output voltage and power of the MCEEH increased and then decreased with the increase of excitation frequency. When the frequency was 8.8 Hz, the MCEEH reached the resonance frequency. The maximum output voltages of coil 1, coil 2, and coil 3 were 1.05 V, 0.8 V, and 0.75 V, respectively. Obviously, the maximum output power of coil 1, coil 2, and coil 3 was 5.2 mW, 2.8 mW, and 2.5 mW, respectively. The MCEEH can effectively harvest energy from low-frequency vibrations. Figure 4 shows the output voltage of coil 1, coil 2, and coil 3 at excitation frequencies of 7 Hz, 8 Hz, 8.8 Hz, 12 Hz, and 16 Hz, respectively. As shown in Figure 4, the output voltages of coil 1, coil 2, and coil 3 were the highest when the resonance frequency was 8.8 Hz. When any magnet vibrated, the other two groups of magnets showed a significant vibration response resulting in a simultaneous voltage output. Therefore, the magnetic coupling method enabled the energy harvester system to harvest energy efficiently at low frequency.

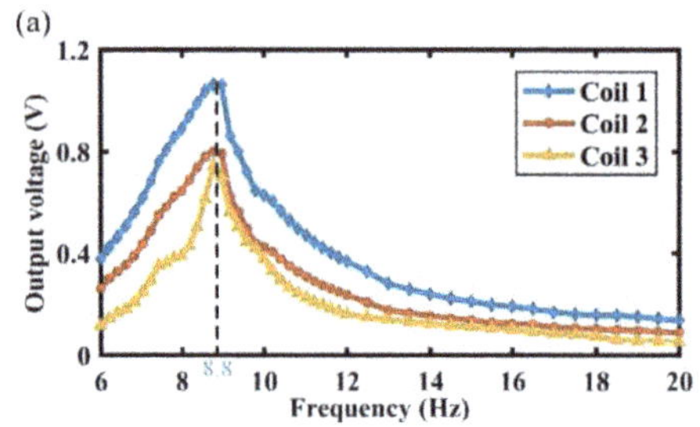
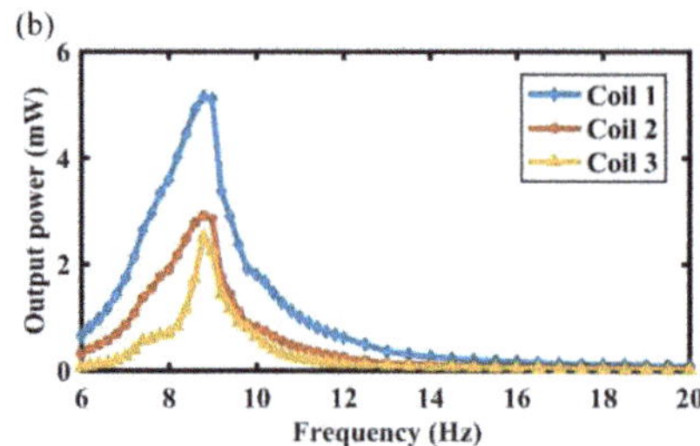

**Figure 3.** The curve of the MEECH output with excitation frequency: (**a**) the curve of the output voltage with the excitation frequency; (**b**) the curve of the output power with the excitation frequency.

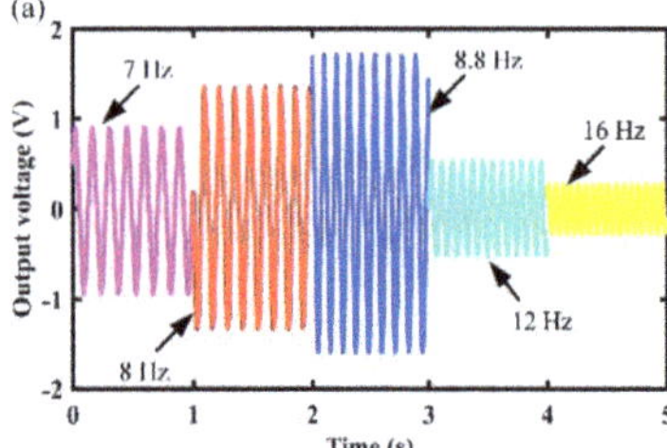
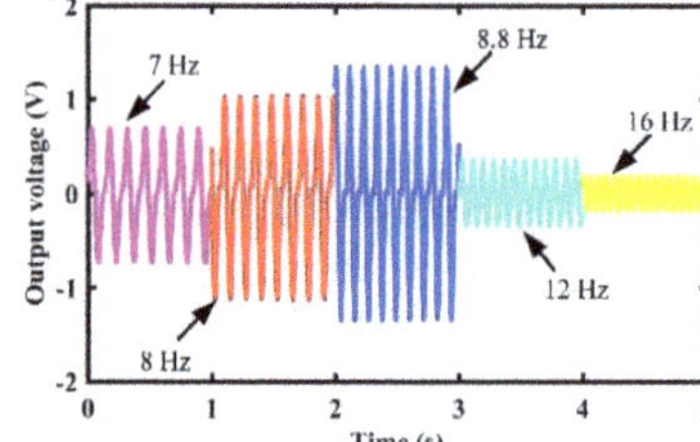
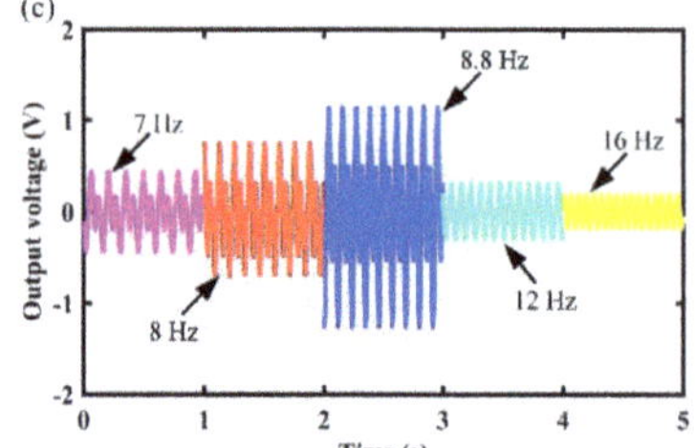

**Figure 4.** Output voltage of the MCEEH at different frequencies: (**a**) coil 1; (**b**) coil 2; (**c**) coil 3.

The output voltage curves of the MCEEH at different excitation accelerations of 0.3 g, 0.5 g, and 0.7 g are shown in Figures 5–7, respectively. It can be found that the resonance frequencies were about 8.6 Hz, 8.8 Hz, and 9.6 Hz at excitation accelerations of 0.3 g, 0.5 g, and 0.7 g, respectively. As can be seen from Figures 5–7, the voltage curves for upward and downward frequency sweep operations were almost identical. Hardening response can be observed in upward and downward sweep operations. Increasing excitation acceleration can enhance the output voltage and improve the energy harvest efficiency at low frequency. The reason for this phenomenon was that the displacement of the center magnet increased and the center magnet can be moved closer to the magnets at both ends, increasing with the excitation acceleration. Enhanced coupling between the magnets increased the output voltage. Repulsion between the center magnet and the magnets at both ends caused the hardening response and the resonance frequency right shift. Nevertheless, the magnetic-spring structure weakened the hardening response. Therefore, there was no significant change in resonance frequency at accelerations of 0.3 g and 0.5 g. As the excitation acceleration increased, the role of repulsive coupling between the magnets in the dynamic response became more and more obvious. The hardening response caused by the repulsion of the magnets was highlighted, which resulting in a large resonance frequency at an acceleration of 0.7 g.

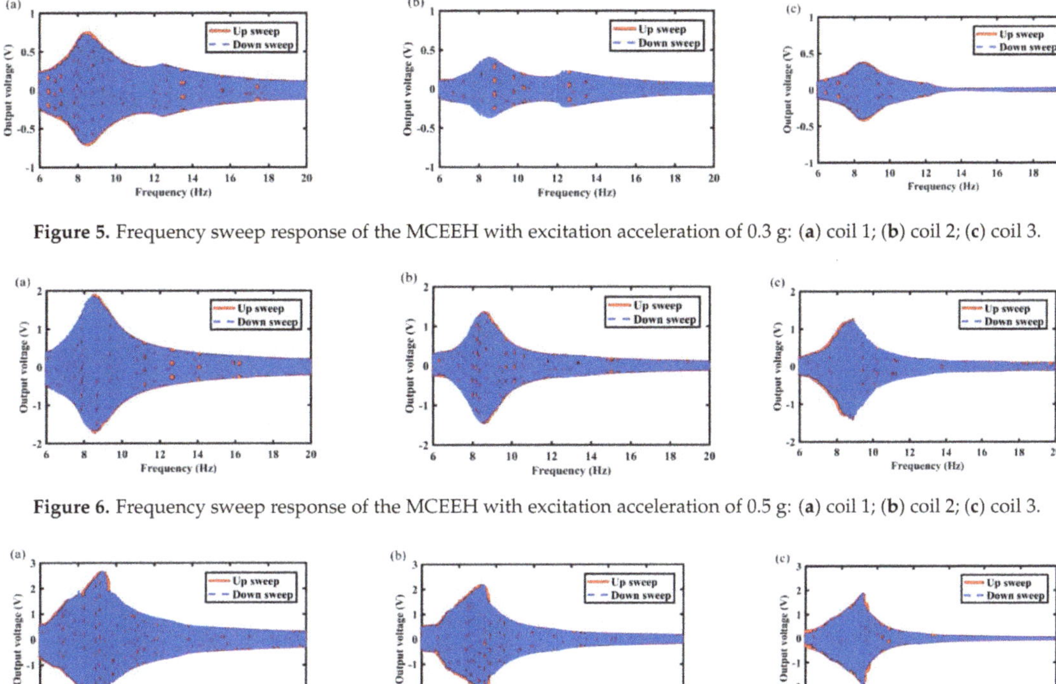

**Figure 5.** Frequency sweep response of the MCEEH with excitation acceleration of 0.3 g: (**a**) coil 1; (**b**) coil 2; (**c**) coil 3.

**Figure 6.** Frequency sweep response of the MCEEH with excitation acceleration of 0.5 g: (**a**) coil 1; (**b**) coil 2; (**c**) coil 3.

**Figure 7.** Frequency sweep response of the MCEEH with excitation acceleration of 0.7 g: (**a**) coil 1; (**b**) coil 2; (**c**) coil 3.

In this section, the effects of spring wire diameter on energy harvest performance are analyzed in detail. For example, Figures 8–10 show the variation of output voltage of the MCEEH versus excitation frequency at spring wire diameters of 0.4 mm, 0.5 mm, and 0.6 mm, respectively. All the excitation accelerations were 0.5 g. The spring wire diameter has a positive correlation with spring stiffness resulting in the resonance frequency of the MCEEH increasing. As displayed in Figure 8, the resonance frequency was 8.8 Hz, and the maximum output voltages of coil 1, coil 2, and coil 3 were 1.8 V, 1.35 V, and 1.2 V, respectively. According to Figure 9, the resonance frequency was 13.3 Hz, and the maximum output voltages of coil 1, coil 2, and coil 3 were 0.66 V, 0.4 V, and 0.72 V, respectively. It is depicted in Figure 10 that the resonance frequency was 15.9 Hz and the maximum output voltages of coil 1, coil 2, and coil 3 were 0.62 V, 0.1 V, and 0.65 V, respectively. It can be seen from Figures 8–10 that with the increase of the spring wire diameter, the output voltages of coil 1, coil 2, and coil 3 all decreased, especially for coil 2. The reason for this phenomenon was that the free vibration of the center magnet was suppressed and the coupling vibration between the center magnet and the magnets at both ends was correspondingly weakened, as the spring stiffness increased. Thus, the output voltage was reduced.

### 3.2. Electrical Output under Hand-Shaking Excitation

In daily activity, the human body can produce kinetic energy in the process of movement. This human body kinetic energy can be harvested by some devices and converted into electrical energy to power portable microelectronic devices. However, human motion is a low frequency and high amplitude vibration, and a free-moving magnet in the MCEEH can improve the efficiency of obtaining energy from human motion. To verify the feasibility of the MCEEH for powering portable microelectronic devices, the hand-shaking excitation experiment was conducted using the MCEEH. The MCEEH was fixed in the holding device, as shown in Figure 11. During the experiment, the device was held by an adult male. During hand-shaking excitation, the excitation acceleration and excitation frequency were approximately 0.2 g and 3.4 Hz, respectively. The real-time voltage output is shown in Figure 12, the peak voltages generated by coil 1, coil 2, and coil 3 were 0.6 V, 0.4 V, and 0.4 V, respectively.

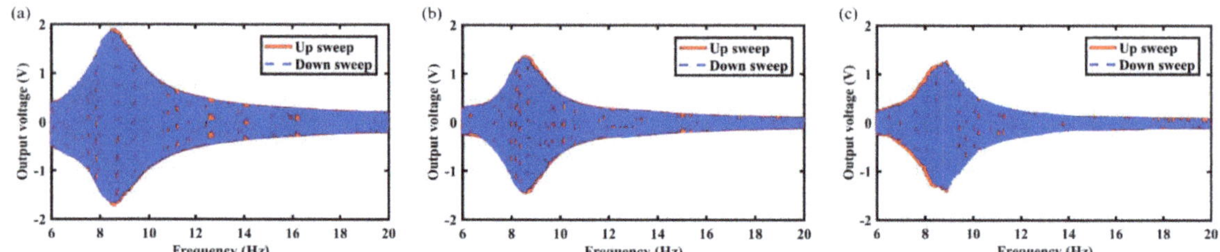

**Figure 8.** Frequency sweep response of the MCEEH with spring wire diameter of 0.4 mm: (**a**) coil 1; (**b**) coil 2; (**c**) coil 3.

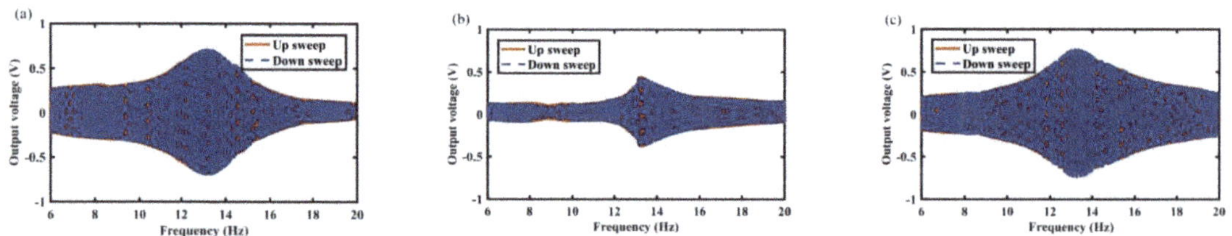

**Figure 9.** Frequency sweep response of the MCEEH with spring wire diameter of 0.5 mm: (**a**) coil 1; (**b**) coil 2; (**c**) coil 3.

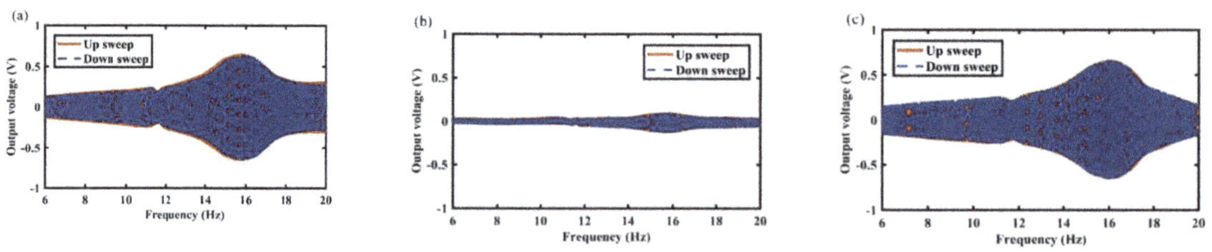

**Figure 10.** Frequency sweep response of the MCEEH with spring wire diameter of 0.6 mm: (**a**) coil 1; (**b**) coil 2; (**c**) coil 3.

### 3.3. Electrical Output under Human Motion Excitation

The MCEEH device was installed on the human body to perform human excitation energy harvest experiments. During the experiment, the energy harvest device was installed on the lower extremity of an adult male with a height of 180 cm. The installation was divided into two: vertical to the leg and parallel to the leg, as shown in Figure 13. In the experiment, the experimenter moved at a constant speed of 3 km/h and 6 km/h, and the voltage generated during the movement was harvested.

Figure 14 shows the output voltage and instantaneous power output response of the MCEEH installed parallel to the leg, respectively. As shown in Figure 14a,b, when the moving speed was 3 km/h, the maximum voltage and maximum instantaneous power reached 0.8 V and 2.7 mW, respectively. It also can be seen from Figure 14c,d, when the moving speed was 6 km/h, the maximum voltage and maximum instantaneous power reached 1.1 V and 4.8 mW, respectively. The voltage had multiple peaks. One peak voltage was generated when the feet touched the ground, and the other peak voltage was generated when the feet were lifted from the ground. With the increase of moving speed, the change of the voltage peak was not obvious. However, the voltage period was significantly reduced. This demonstrated that the MCEEH can effectively harvest energy in high-speed motion in the state of parallel installation.

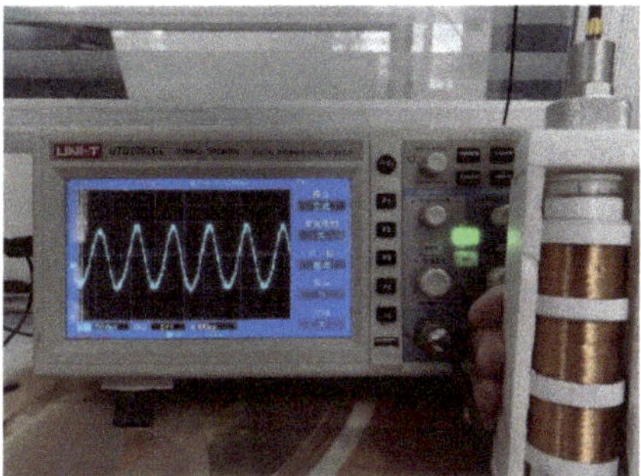

**Figure 11.** Picture of the hand-shaking test.

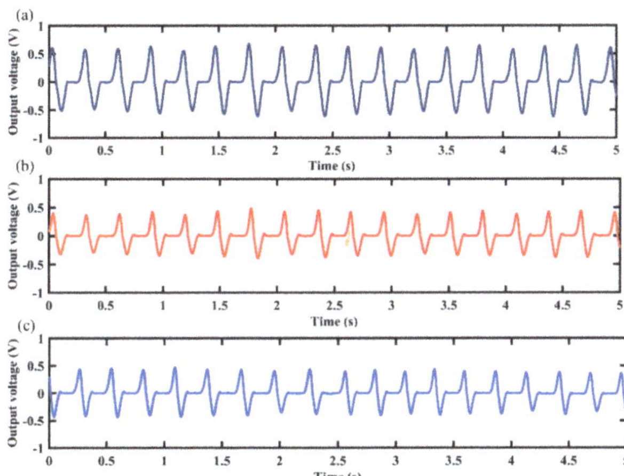

**Figure 12.** Electrical output under hand-held excitation: (**a**) output voltage of coil 1; (**b**) output voltage of coil 2; (**c**) output voltage of coil 3.

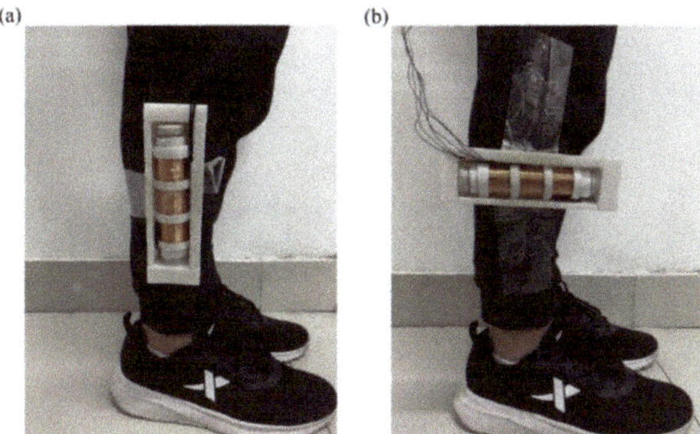

**Figure 13.** Picture of the human motion excitation test: (**a**) parallel to the leg; (**b**) vertical to the leg.

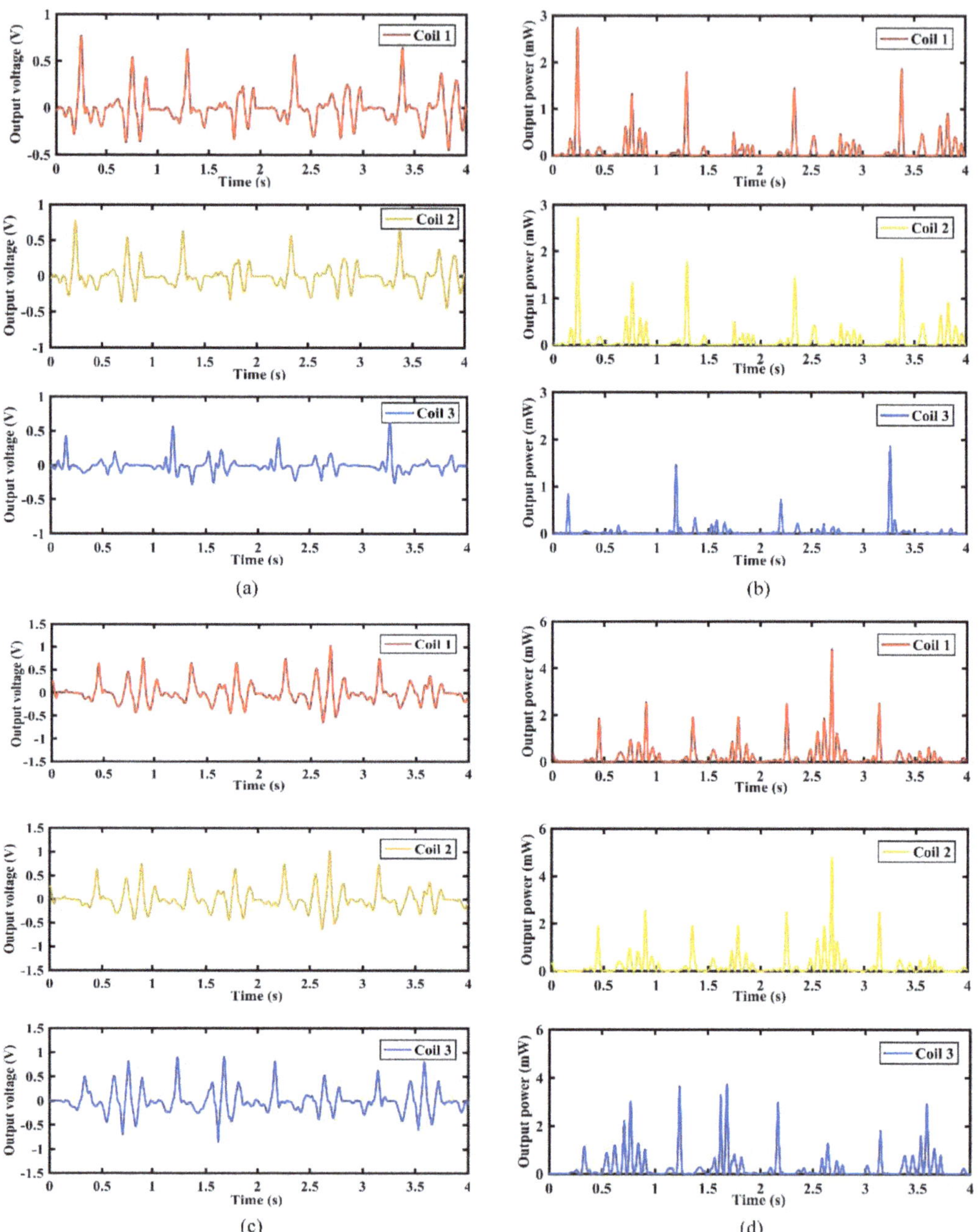

**Figure 14.** Output of the MCEEH installed parallel to the leg: (**a**) output voltage for moving speed at 3 km/h; (**b**) instantaneous power output for moving speed at 3 km/h; (**c**) output voltage for moving speed at 6 km/h; (**d**) instantaneous power output for moving speed at 6 km/h.

Figure 15 shows the output voltage and instantaneous power output response of the MCEEH installed vertically to the leg. As shown in Figure 15a,b, when the moving speed was 3 km/h, the maximum voltage and maximum instantaneous power reached

0.8 V and 7.4 mW, respectively. It also can be seen from Figure 15c,d, when the moving speed was 6 km/h, the maximum voltage and maximum instantaneous power reached 2.4 V and 26 mW, respectively. The step frequency of 3 km/h was about 0.9 Hz, and the step frequency of 6 km/h was about 1.4 Hz. The voltage increased with higher moving speed. Under the condition of large step frequency, the output voltage was large. This showed that the MCEEH can effectively harvest energy in low frequency in the state of vertical installation.

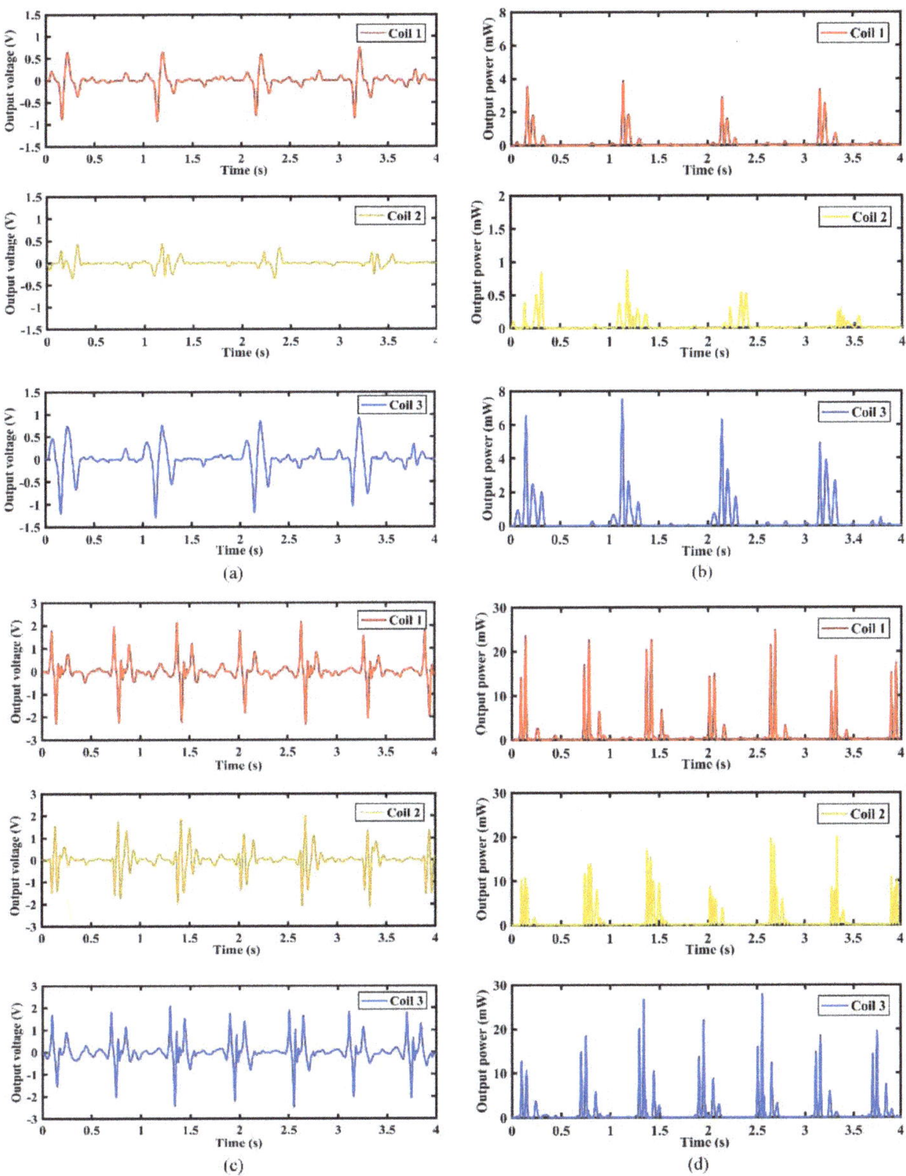

**Figure 15.** Output of the MCEEH installed vertical to the leg: (**a**) output voltage for moving speed at 3 km/h; (**b**) instantaneous power output for moving speed at 3 km/h; (**c**) output voltage for moving speed at 6 km/h; (**d**) instantaneous power output for moving speed at 6 km/h.

To verify the performance of the MCEEH without affecting human motion, the MCEEH was placed vertically in a backpack, as shown in Figure 16a. The experimenter traveled with the backpack on a treadmill with constant speeds of 3 km/h, 4 km/h, 5 km/h, 6 km/h, and 7 km/h, respectively. The output power of the MCEEH at different speeds is shown in Figure 16b. The maximum power values of coil 1, coil 2, and coil 3 were 1.5 mW, 2.1 mW, and 1.4 mW, respectively. It is clear from Figure 16 that the output power of the MCEEH increased slightly when the speed varied from 3 to 5 km/h. However, the output power of the MCEEH increased rapidly as the experimenter's speed changed from 6 to 7 km/h. The reason for this phenomenon was the change in the movement of the experimenter from walking to running.

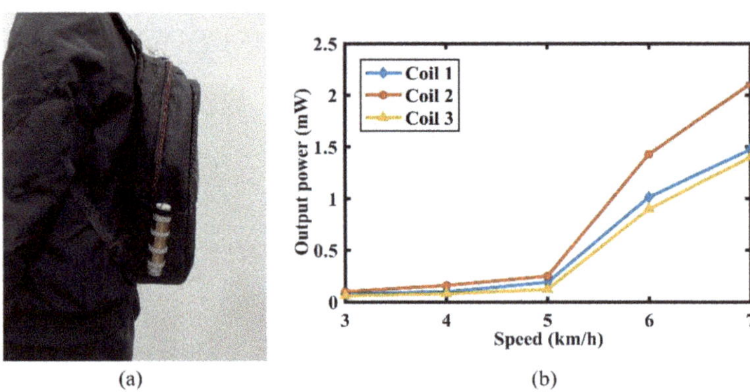


(a)
(b)

**Figure 16.** (**a**) Picture of the MCEEH put in a backpack vertically; (**b**) the output power of MCEEH at different speeds.

## 4. Conclusions

In this paper, a magnetically coupled electromagnetic energy harvester (MCEEH) is proposed for converting human body kinetic energy into electric energy. The main novel structure of the MCEEH consists of a pair of spring-connected magnets, coils, and a free-moving magnet. Specifically, the interaction force between the magnets is repulsive. This structure shows a typical hardening response to harmonic excitation. The magnetic-spring structure can weaken the hardening response caused by the repulsive force. A free-moving magnet in the MCEEH can improve the efficiency of obtaining energy from human motion. The effects of excitation frequency, excitation acceleration, and spring stiffness on the vibration response and output voltage of the MCEEH have been experimentally studied. In detail, increasing the excitation acceleration can improve the output voltage. The MCEEH has the advantage of adjustable working frequency and high low-frequency energy harvest efficiency. Increasing the spring stiffness causes the resonance frequency to shift to the right and the output voltage to decrease. The human body kinetic energy experiment was conducted using an MCEEH. For example, the generator can obtain the maximum voltage of 0.6 V during hand-shaking excitation. Furthermore, maximum instantaneous power can be obtained of about 4.8 mW and 26 mW in the case of parallel and vertical orientation to the leg. Maximum power of 2.1 mW can be obtained when placed vertically in the backpack. If the performance of the harvester is further improved, such as the power can be increased by increasing the number of coil turns and using thicker magnets, the MCEEH can be used to power various microelectronic products, e.g., wrist watches, sensors, hearing aids, etc. Accordingly, this paper provides a feasible guideline to improve performance for electromagnetic energy harvesters.

**Author Contributions:** X.L., the whole work including design, fabrication, testing of the device, and writing—original draft preparation; J.M., testing guidance; C.Y., simulation analysis and writing—review and editing; H.Z., fabrication and simulation analysis; L.Z., writing—review and editing; R.S., research idea and gave guidance. All authors have read and agreed to the published version of the manuscript.

**Funding:** This work was supported by the National Natural Science Foundation of China under grant numbers 51705296, 52075305 and 52075306; Shenzhen Science and Technology Innovation Committee under grant No. JCYJ20200109143206663; and the Key Research and Development Project of Zibo City under grant No. 2020SNPT0088.

**Acknowledgments:** The authors are also grateful to the experimental support of Shandong University of Technology.

**Conflicts of Interest:** The authors declare no conflict of interest.

# References

1. Glynne-Jones, P.; Tudor, M.J.; Beeby, S.P.; White, N.M. An electromagnetic, vibration-powered generator for intelligent sensor systems. *Sens. Actuators A* **2004**, *110*, 344–349. [CrossRef]
2. Erturk, A.; Renno, J.M.; Inman, D.J. Modeling of Piezoelectric Energy Harvesting from an L-shaped Beam-mass Structure with an Application to UAVs. *J. Intell. Mater. Syst. Struct.* **2008**, *20*, 529–544. [CrossRef]
3. Kecik, K.; Mitura, A.; Lenci, S.; Warminski, J. Energy harvesting from a magnetic levitation system. *Int. J. Non Linear Mech.* **2017**, *94*, 200–206. [CrossRef]
4. Stephen, N.G. On energy harvesting from ambient vibration. *J. Sound Vib.* **2006**, *293*, 409–425. [CrossRef]
5. Kan, J.; Fan, C.; Wang, S.; Zhang, Z.; Wen, J.; Huang, L. Study on a piezo-windmill for energy harvesting. *Renew. Energy* **2016**, *97*, 210–217. [CrossRef]
6. Yang, Z.; Zhou, S.; Zu, J.; Inman, D. High-Performance Piezoelectric Energy Harvesters and Their Applications. *Joule* **2018**, *2*, 642–697. [CrossRef]
7. Xie, X.D.; Wang, Q.; Wu, N. A ring piezoelectric energy harvester excited by magnetic forces. *Int. J. Eng. Sci.* **2014**, *77*, 71–78. [CrossRef]
8. Pillatsch, P.; Yeatman, E.M.; Holmes, A.S. A piezoelectric frequency up-converting energy harvester with rotating proof mass for human body applications. *Sens. Actuators A* **2014**, *206*, 178–185. [CrossRef]
9. Toyabur, R.M.; Salauddin, M.; Cho, H.; Park, J.Y. A multimodal hybrid energy harvester based on piezoelectric-electromagnetic mechanisms for low-frequency ambient vibrations. *Energy Convers. Manag.* **2018**, *168*, 454–466. [CrossRef]
10. Dai, H.L.; Abdelkefi, A.; Javed, U.; Wang, L. Modeling and performance of electromagnetic energy harvesting from galloping oscillations. *Smart Mater. Struct.* **2015**, *24*, 045012. [CrossRef]
11. Liu, R.; Xu, Z.; Jin, Y.; Wang, W. Design and research on a nonlinear 2dof electromagnetic energy harvester with velocity amplification. *IEEE Access* **2020**, *8*, 159947–159955. [CrossRef]
12. Halim, M.A.; Rantz, R.; Zhang, Q.; Gu, L.; Yang, K.; Roundy, S. An electromagnetic rotational energy harvester using sprung eccentric rotor, driven by pseudo-walking motion. *Appl. Energy* **2018**, *217*, 66–74. [CrossRef]
13. Zhang, L.B.; Dai, H.L.; Yang, Y.W.; Wang, L. Design of high-efficiency electromagnetic energy harvester based on a rolling magnet. *Energy Convers. Manag.* **2019**, *185*, 202–210. [CrossRef]
14. Zhang, H.; Sui, W.; Yang, C.; Zhang, L.; Song, R.; Yang, X. Scavenging wind induced vibration by an electromagnet energy harvester from single to multiple wind directions. *Ferroelectrics* **2021**, *577*, 170–180. [CrossRef]
15. Zhang, Y.; Wang, T.; Luo, A.; Hu, Y.; Li, X.; Wang, F. Micro electrostatic energy harvester with both broad bandwidth and high normalized power density. *Appl. Energy* **2018**, *212*, 362–371. [CrossRef]
16. Dragunov, V.P.; Ostertak, D.I.; Sinitskiy, R.E. New modifications of a Bennet doubler circuit-based electrostatic vibrational energy harvester. *Sens. Actuators A* **2020**, *302*, 111812. [CrossRef]
17. Murotani, K.; Suzuki, Y. MEMS electret energy harvester with embedded bistable electrostatic spring for broadband response. *J. Micromech. Microeng.* **2018**, *28*, 104001. [CrossRef]
18. Lu, Y.; Capo-Chichi, M.; Leprince-Wang, Y.; Basset, P. A flexible electrostatic kinetic energy harvester based on electret films of electrospun nanofibers. *Smart Mater. Struct.* **2018**, *27*, 14001. [CrossRef]
19. Foong, F.M.; Thein, C.K.; Yurchenko, D. Important considerations in optimising the structural aspect of a SDOF electromagnetic vibration energy harvester. *J. Sound Vib.* **2020**, *482*, 115470. [CrossRef]
20. Halim, M.A.; Park, J.Y. Modeling and experiment of a handy motion driven, frequency up-converting electromagnetic energy harvester using transverse impact by spherical ball. *Sens. Actuators A* **2015**, *229*, 50–58. [CrossRef]
21. Fang, S.; Zhou, S.; Yurchenko, D.; Yang, T.; Liao, W.-H. Multistability phenomenon in signal processing, energy harvesting, composite structures, and metamaterials: A review. *Mech. Syst. Signal Process.* **2022**, *166*, 108419. [CrossRef]
22. Huang, D.; Chen, J.; Zhou, S.; Fang, X.; Li, W. Response regimes of nonlinear energy harvesters with a resistor-inductor resonant circuit by complexification-averaging method. *Sci. China Technol. Sci.* **2021**, *64*, 1212–1227. [CrossRef]

23. Kuroki, J.; Shinshi, T.; Li, L.; Shimokohbe, A. Miniaturization of a one-axis-controlled magnetic bearing. *Precis. Eng.* **2005**, *29*, 208–218. [CrossRef]
24. Saha, C.R.; O'Donnell, T.; Wang, N.; McCloskey, P. Electromagnetic generator for harvesting energy from human motion. *Sens. Actuators A* **2008**, *147*, 248–253. [CrossRef]
25. Mann, B.P.; Sims, N.D. Energy harvesting from the nonlinear oscillations of magnetic levitation. *J. Sound Vib.* **2009**, *319*, 515–530. [CrossRef]
26. Mann, B.P.; Owens, B.A. Investigations of a nonlinear energy harvester with a bistable potential well. *J. Sound Vib.* **2010**, *329*, 1215–1226. [CrossRef]
27. Lee, B.-C.; Rahman, M.A.; Hyun, S.-H.; Chung, G.-S. Low frequency driven electromagnetic energy harvester for self-powered system. *Smart Mater. Struct.* **2012**, *21*, 125024. [CrossRef]
28. Foisal, A.R.M.; Hong, C.; Chung, G.-S. Multi-frequency electromagnetic energy harvester using a magnetic spring cantilever. *Sens. Actuators A* **2012**, *182*, 106–113. [CrossRef]
29. Fan, K.; Cai, M.; Liu, H.; Zhang, Y. Capturing energy from ultra-low frequency vibrations and human motion through a monostable electromagnetic energy harvester. *Energy* **2019**, *169*, 356–368. [CrossRef]
30. Fan, K.; Zhang, Y.; Liu, H.; Cai, M.; Tan, Q. A nonlinear two-degree-of-freedom electromagnetic energy harvester for ultra-low frequency vibrations and human body motions. *Renew. Energy* **2019**, *138*, 292–302. [CrossRef]
31. Zhu, D.; Beeby, S.P. A broadband electromagnetic energy harvester with a coupled bistable structure. *J. Phys. Conf. Ser.* **2013**, *476*, 012070. [CrossRef]
32. Masana, R.; Daqaq, M.F. Relative performance of a vibratory energy harvester in mono- and bi-stable potentials. *J. Sound Vib.* **2011**, *330*, 6036–6052. [CrossRef]
33. Munaz, A.; Lee, B.-C.; Chung, G.-S. A study of an electromagnetic energy harvester using multi-pole magnet. *Sens. Actuators A* **2013**, *201*, 134–140. [CrossRef]
34. Halim, M.A.; Cho, H.; Park, J.Y. Design and experiment of a human-limb driven, frequency up-converted electromagnetic energy harvester. *Energy Convers. Manag.* **2015**, *106*, 393–404. [CrossRef]

*micromachines*

*Article*

# Levitation Characteristics Analysis of a Diamagnetically Stabilized Levitation Structure

Shuhan Cheng, Xia Li, Yongkun Wang and Yufeng Su *

School of Mechanical and Power Engineering, Zhengzhou University, Zhengzhou 450001, China; ritchie1028@163.com (S.C.); Lx2007@zzu.edu.cn (X.L.); Wangyongkun1999@163.com (Y.W.)
* Correspondence: yufengsu@zzu.edu.cn

**Abstract:** A diamagnetically stabilized levitation structure is composed of a floating magnet, diamagnetic material, and a lifting magnet. The floating magnet is freely levitated between two diamagnetic plates without any external energy input. In this paper, the levitation characteristics of a floating magnet were firstly studied through simulation. Three different levitation states were found by adjusting the gap between the two diamagnetic plates, namely symmetric monostable levitation, bistable levitation, and asymmetric monostable levitation. Then, according to experimental comparison, it was found that the stability of the symmetric monostable levitation system is better than that of the other two. Lastly, the maximum moving space that allows the symmetric monostable levitation state is investigated by Taguchi method. The key factors affecting the maximum gap were determined as the structure parameters of the floating magnet and the thickness of highly oriented pyrolytic graphite (HOPG) sheets. According to the optimal parameters, work performance was obtained by an experiment with an energy harvester based on the diamagnetic levitation structure. The effective value of voltage is 250.69 mV and the power is 86.8 μW. An LED light is successfully lit on when the output voltage is boosted with a Cockcroft–Walton cascade voltage doubler circuit. This work offers an effective method to choose appropriate parameters for a diamagnetically stabilized levitation structure.

**Keywords:** diamagnetically stabilized levitation; Taguchi method; stable levitation; maximum gap

**Citation:** Cheng, S.; Li, X.; Wang, Y.; Su, Y. Levitation Characteristics Analysis of a Diamagnetically Stabilized Levitation Structure. *Micromachines* **2021**, *12*, 982. https://doi.org/10.3390/mi12080982

Academic Editors: Qiongfeng Shi and Huicong Liu

Received: 9 June 2021
Accepted: 16 August 2021
Published: 19 August 2021

## 1. Introduction

Diamagnetism is a natural property of a substance and exists in all materials. However, it is not easily appreciable in daily life, because it is too weak, compared to magnetism and paramagnetism. To observe diamagnetism, the diamagnetic material needs to be placed in a strong external magnetic field [1]. In an external magnetic field, the diamagnetic material generates a weak magnetic field, which is opposite to the external magnetic field. As a result, the diamagnetic material is subjected to a repelling force from the external magnetic field. When the repelling force and gravity of the diamagnetic material are equal and opposite to each other, the diamagnetic material is levitated in the external magnetic field, which is known as diamagnetic levitation. In 1939, diamagnetic levitation [2] was first observed by experiments where a small piece of bismuth and graphite was freely levitated in a strong electromagnetic field. In 2000, Simon et al. [3] further studied diamagnetic levitation and proposed diamagnetically stabilized levitation, which is a variant of diamagnetic levitation. In the study, the magnet served as a floater and was stably levitated between the diamagnetic materials without any external energy input.

In recent years, some applications based on diamagnetically stabilized levitation have been reported, such as sensors [4–6], actuators [7], and vibration energy harvesters [8–10]. Hilber et al. [11] presented a sensor based on diamagnetically stabilized levitation, which can be used to measure the density and viscosity of fluids in microfluidic systems. Ye et al. [12] designed a vibration energy harvester using diamagnetically stabilized levitation to harvest ambient vibration energy. Ding et al. [13] from the same research group conducted simulation and analysis on the energy harvester, and conducted experimental tests on a set of structural parameters. Clara et al. [14] investigated a viscosity and density sensor using diamagnetically stabilized levitation of a floater magnet on pyrolytic graphite. Liu et al. [15] studied a diamagnetically levitated electrostatic micromotor, which were fabricated by MSMS process and precision machining. Gisela et al. [16] constructed a low-cost magnetic levitation system. Chow et al. [17] studied the shape effect of magnetic sources formed by standard coil and ring magnet elements on diamagnetically stabilized levitation.

In this paper, by studying the static levitation characteristics of a structure constructed by Ding et al. [13], the levitation characteristics of the floating magnet in diamagnetically stabilized levitation are analyzed by simulation and experiments, and it was found that the floating magnet has three different levitation states, namely symmetric monostable levitation, bistable levitation, and asymmetric monostable levitation. Three levitation states were obtained by adjusting the gap of the diamagnetic materials. In order to make the energy harvester have better output characteristics, the moving space of the floating magnet is introduced, and the increase of this parameter is conducive to the arrangement of more coils. The maximum moving space that allows the floating magnet to achieve symmetric monostable levitation is determined by the structure parameters of the diamagnetically stabilized levitation. The influence of the structure parameters on the maximum moving space was studied by the Taguchi method. It was found through experiments that this method can effectively optimize the selection of structural parameters and improve the output characteristics.

## 2. Theory of Diamagnetically Stabilized Levitation

The structure of the diamagnetically stabilized levitation is shown in Figure 1a, which consists of a lifting magnet, an upper HOPG sheet, a floating magnet, and a lower HOPG sheet. The floating magnet is stably levitated between the two HOPG sheets.

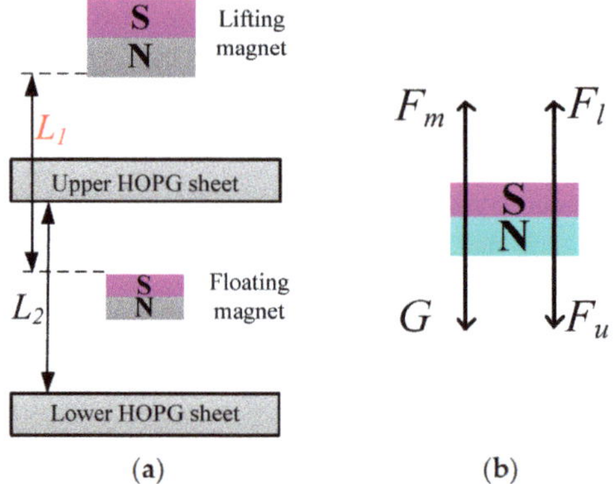


**Figure 1.** (**a**) Schematic of the diamagnetically stabilized levitation; (**b**) Mechanic analysis of the floating magnet.

The potential energy of the floating magnet in the field of the lifting magnet can be written as follows [3]:

$$U = -\vec{M} \cdot \vec{B} + mgz = -MB + mgz \tag{1}$$

where $\vec{M}$ and $m$ are the magnetic dipole moment and mass of the floating magnet, respectively, $g$ is the gravity acceleration, $B$ is the magnetic flux density of the lifting magnet, $z$ is the distance of the magnet orthogonal to the reference surface. With magnetic torques, the floating magnet aligns with the local field direction. As a result, energy only relies on the magnitude of the magnetic field.

Expanding the field magnitude of the lifting magnet around the levitation position in polar coordinates and adding two new terms $C_z z^2$ and $C_r r^2$ which denote the effect of diamagnetic materials, the potential energy of the floating magnet can be rewritten as:

$$U = -M\left[B_0 + \left\{B' - \frac{mg}{M}\right\}z + \frac{1}{2}B''z^2 + \frac{1}{4}\left\{\frac{B'^2}{2B_0} - B''\right\}r^2 + \cdots\right] + C_z z^2 + C_r r^2 \quad (2)$$

where $B' = \frac{\partial B_z}{\partial z}$ and $B'' = \frac{\partial^2 B_z}{\partial z^2}$.

The expression in the first curly bracket must be equal to zero when the floating magnet locates at the levitation position. In other words, the gravity of the floating magnet is balanced by the force derived from the non-uniform magnetic field:

$$B' = \frac{mg}{M} \quad (3)$$

Furthermore, the conditions for vertical stability and horizontal stability can be derived according to Equation (2):

$$K_v \equiv C_z - \frac{1}{2}MB'' > 0 \text{ Vertical stability} \quad (4)$$

$$K_h \equiv C_r + \frac{1}{4}M\left\{B'' - \frac{B'^2}{2B_0}\right\} = C_r + \frac{1}{4}M\left\{B'' - \frac{m^2g^2}{2M^2B_0}\right\} > 0 \text{ Horizontal stability} \quad (5)$$

To achieve a stable levitation for the floating magnet, these conditions are necessary to ensure a local minimum of $U$ at the equilibrium point. When Equations (4) and (5) are fulfilled, the stable levitation is possible if $MB' = mg$. Therefore, the condition can be matched by adjusting the field gradient or the weight of the floating magnet.

In addition, the energy generated by two HOPG sheets can be expressed as [18]:

$$U_{dia} = C_z z^2 = \frac{6\mu_0 M^2 |\chi|}{\pi L_2{}^5}z^2 \quad (6)$$

where $L_2$ is the gap between two HOPG sheets, $\chi$ is the magnetic susceptibility of the diamagnetic material.

According to Equations (4)–(6), the condition of the stable levitation can be obtained at the point where $B' = mg/M$, which can be written as follows:

$$\frac{12\mu_0 M |\chi|}{\pi L_2{}^5} > B'' > \frac{(mg)^2}{2M^2 B_0} \quad (7)$$

This puts a limit on the gap $L_2$ [17]:

$$L_2 < \left\{\frac{12\mu_0 M |\chi|}{\pi B''}\right\}^{1/5} < \left\{\frac{24\mu_0 B_0 M^3 |\chi|}{\pi (mg)^2}\right\}^{\frac{1}{5}} \quad (8)$$

It can be seen that the gap $L_2$ should be limited in a certain range for stabilizing the floating magnet. However, the levitation characteristic of a floating magnet has not been discussed with different gap $L_2$.

## 3. Analysis of Levitation Characteristics

To understand the levitation characteristic, the mechanics analysis of the floating magnet is performed, which is shown in Figure 1b. Since the magnetization directions of the two magnets are the same, an upward magnetic traction $F_m$ is exerted on the floating magnet by the lifting magnet. In addition, two opposite repulsive forces ($F_u$ and $F_l$)

generated by two HOPG sheets simultaneously act on the floating magnet. Therefore, the resultant force $F_r$ exerted on the floating magnet can be written as follows:

$$F_r = F_m + F_l - F_u - G \tag{9}$$

where $G$ is the gravity of the floating magnet.

When the floating magnet is levitated at an equilibrium position, the resultant force is equal to zero. As shown in Figure 2, finite element analysis (FEA) simulation was performed by COMSOL Multiphysics 5.5, so as to obtain the resultant force. The structure parameters used in the simulation are listed in Table 1, and the simulation results are shown in Figure 3. In the analysis, the symmetrical plane of two HOPG sheets is selected as zero-plane, and the upward direction is set as positive. When $L_2$ is less than 6.2 mm, the resultant force curve has only one point. The numbers of zero point are increased to three when the gap $L_2$ is in the range of 6.2–7.0 mm. There are two zero points when $L_2$ is equal to 7.0 mm. Zero resultant force indicates that the floating magnet can achieve an equilibrium state at these positions, but it does not mean that the floating magnet can realize a stable levitation. The levitation characteristic of the floating magnet cannot be exactly determined by the resultant force, which needs to refer to the potential energy of the floating magnet.

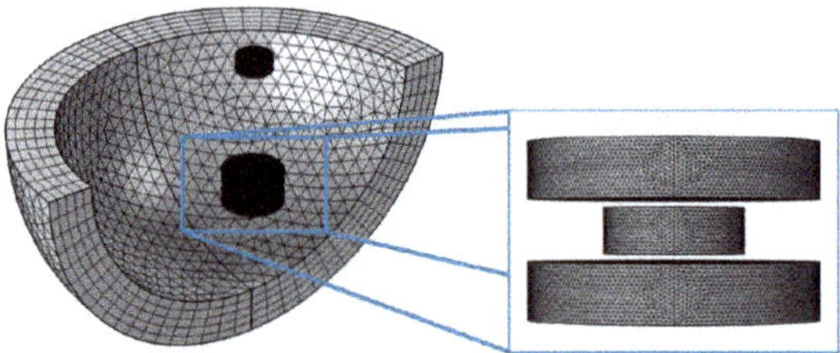

**Figure 2.** FEA model of the diamagnetically stabilized levitation.

**Table 1.** Structure parameters of the diamagnetically stabilized levitation.

| Parameter | Value/Material |
| --- | --- |
| Lifting magnet and floating magnet | NdFeB-52 |
| Residual flux density of floating magnet(T) | 1.45 |
| Residual flux density of lifting magnet(T) | 1.45 |
| Size of lifting magnet(mm) | $\Phi\ 15 \times 6.35$ |
| Size of floating magnet(mm) | $\Phi\ 12 \times 4$ |
| Size of diamagnetic sheets(mm) | $\Phi\ 25 \times 5$ |
| Material of diamagnetic sheets | HOPG |
| Density of floating magnet (kg/m$^3$) | $7.5 \times 10^3$ |
| Magnetic susceptibility $\chi$ of HOPG | $[8, 8, 45] \times 10^{-5}$ |

Figure 4 shows the potential energy of the floating magnet in the cases shown in Figure 3. According to the principle of minimum potential energy, a system will be in a stable equilibrium state when its potential energy reaches a local minimum. For the diamagnetically stabilized levitation structure, the local minimum of the potential energy does not always occur at these positions where the resultant force is equal to zero. Therefore, the floating magnet cannot be stably levitated at all the positions with zero resultant force. When $L_2$ is less than 6.2 mm, the potential energy curve has only one local minimum, which means the floating magnet can only be stably levitated at one position. Moreover, the stable levitation position is in the zero-plane, and this state is named symmetric monostable levitation. When $L_2$ is equal to 6.6 mm, two different minimum points appear on the potential energy curve, which indicates the floating magnet has two different stable levitation positions. In addition, two minimum points are not in the zero-plane. In the zero-plane, the floating magnet can also reach an equilibrium state because of the zero resultant force. However, this equilibrium state is easily broken by a slight external disturbance, which leads to a non-stable equilibrium. The feature with two stable levitation points is also known as bistable levitation. Adjusting $L_2$ to 7.0 mm, the stable levitation point above zero-plane will disappear due to the large gradient of the magnetic field near the lifting magnet. In this case, the floating magnet can only be levitated below zero-plane. The phenomenon is termed asymmetric monostable levitation.

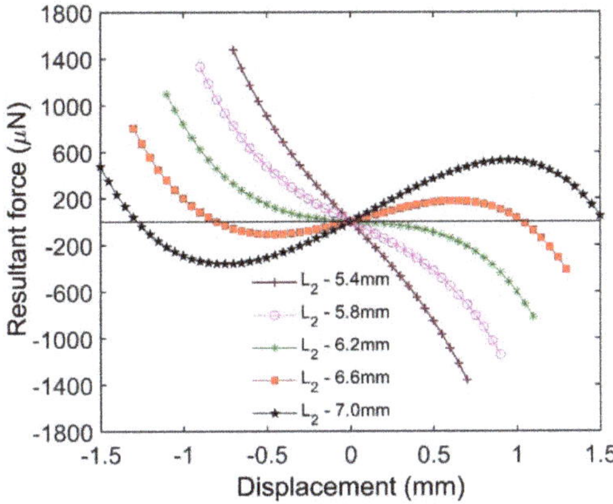

**Figure 3.** Resultant force with respect to various displacements for different gaps $L_2$.

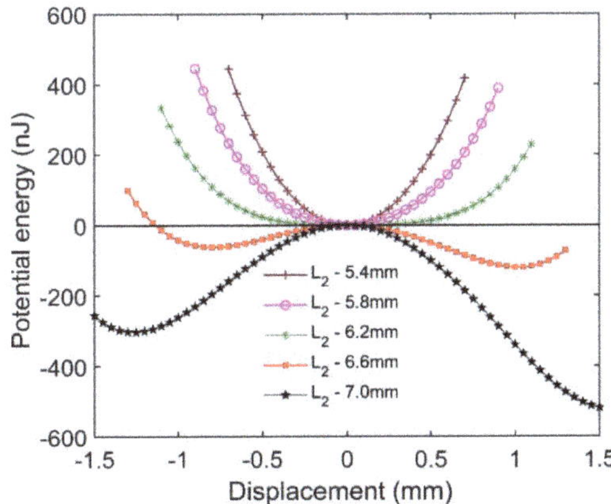

**Figure 4.** Potential energy with respect to various displacements for different gaps $L_2$.

To verify the simulation results, an experimental setup was put up, which is shown in Figure 5. Two support sheets are mounted on two precision adjustment tables installed on an aluminum plate, which is used to support the lifting magnet and the upper HOPG sheet. The lower HOPG sheet is directly fixed on the aluminum plate, the upper HOPG sheet is mounted on the lower face of the support sheet, and the lifting magnet is located on the upper face of the other support sheet. $L_1$ and $L_2$ can be adjusted by two precision adjustment tables. $L_1$ is the distance between the lifting magnet and the floating magnet.

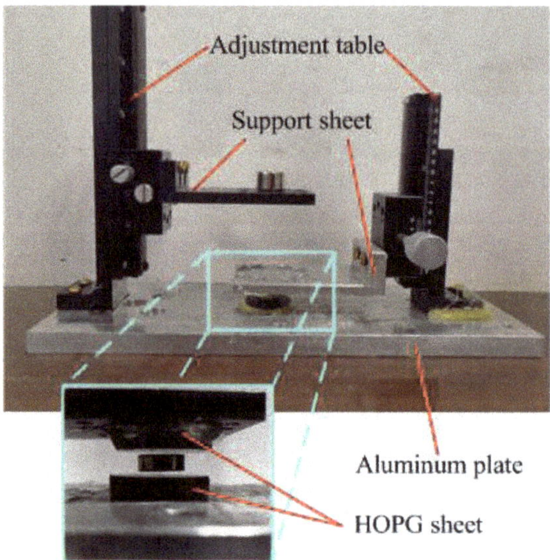



**Figure 5.** Experiment setup.

Three different levitation states were verified one by one by adjusting the gap $L_2$ between two HOPG sheets, which are shown in Figure 6. In the symmetric monostable levitation state, the floating magnet can always return to the initial position under an impact excitation. When an impact excitation is applied to the system with a bistable levitation state, the floating magnet may jump between the two equilibrium points and eventually stop at one point. In the asymmetric monostable levitation state, the floating magnet will vibrate near the equilibrium point when a slight impact excitation is adopted. Increasing the intensity of the external excitation, the floating magnet will pass through the zero-plane and be firmly adsorbed on the upper HOPG sheet due to the magnetic traction Fm. Among the three levitation states, the stability of the symmetric monostable levitation state is the best. Hence, the structure of the symmetric monostable levitation state is more suitable for developing new applications, such as sensors, actuators, and vibration energy harvesters.

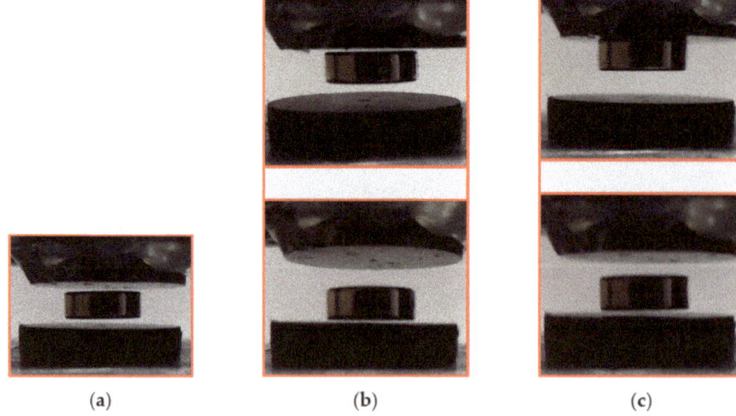

(a)           (b)           (c)

**Figure 6.** (**a**) Symmetric monostable levitation state; (**b**) bistable levitation state; (**c**) asymmetric monostable levitation state.

## 4. Analysis of Maximum Moving Space $\sigma$

In the symmetric monostable structure, the maximum of the gap $L_2$ is a key parameter. The maximum moving space $\sigma$ derived from $L_2$ is an important indicator of structural performance. It represents the movable space of the floating magnet in the vertical direction. It is numerically equal to the subtraction between $L_2$ and the thickness of the floating magnet. The maximum moving space $\sigma$ is determined by the part parameters, which include coating thickness, residual flux density, and structure size. Since the coating thickness and residual flux density are provided by the manufacturer, we just focus on the effect of the structure parameters on $\sigma$. To analyze the influence of structure parameters, the Taguchi method [19–23] is an excellent tool and is adopted in this analysis. The orthogonal array is used in the Taguchi method to arrange an experiment, which is composed of factors and levels. The experiment results are classified into three different categories by signal-to-noise (S/N) ratio: the larger-the-better (LB), the nominal-the-better (NB), and the smaller-the-better (SB). In the S/N ratio, the signal represents the desired value, whereas noise represents the undesired measured value. The S/N ratio is a parameter that can be used to evaluate the sensitivity of a parameter on the physical behavior, which is widely used to indicate the engineering quality. A larger S/N ratio corresponds to a better quality of a system. In this analysis, the objective is to maximize moving space $\sigma$ of the symmetric monostable system; the LB criterion is adopted. The S/N ratio in terms of the maximum moving space $\sigma$ is expressed as:

$$\frac{S}{N} = -10\log_{10}\left(\frac{1}{y^2}\right) \tag{10}$$

where $y$ is the maximum value of moving space $\sigma$.

Seven control factors, consisting of the thickness, inner diameter, and outer diameter of the lifting magnet; the thickness, inner diameter, and outer diameter of the floating magnet; and the thickness of the HOPG sheets, along with three levels, are taken into account. These structural parameters and corresponding levels are listed in Table 2. For a Taguchi approach with 7 factors and 3 levels, a typical orthogonal array $L_{27}$ ($3^7$) with 27 runs is given in Table 3. The maximum value of the moving space $\sigma$ and corresponding S/N ratios are listed in Table 3.

**Table 2.** Structure parameters and their levels.

| Symbol | Factor (mm) | Level 1 | Level 2 | Level 3 |
|:---:|:---:|:---:|:---:|:---:|
| A | Outer diameter of floating magnet | 10 | 12 | 15 |
| B | Inner diameter of floating magnet | 0 | 3.175 | 6.35 |
| C | Thickness of floating magnet | 2 | 4 | 5 |
| D | Outer diameter of lifting magnet | 10 | 12.7 | 15 |
| E | Inner diameter of lifting magnet | 0 | 3.175 | 6.35 |
| F | Thickness of lifting magnet | 3.175 | 6.35 | 10 |
| G | Thickness of HOPG | 1 | 3 | 5 |

### 4.1. Analysis of Variance

The analysis of variance (ANOVA) is performed for moving space $\sigma$, as listed in Table 4, to evaluate the contribution of the factors. The P-magnitude of the control factors declares the statistical significance to the confidence level of 0.95 [19]. The $P$ value infers that the relevant parameters of the lifting magnet have an insignificant effect on the maximum moving space of the floating magnet ($P > 0.05$). The thickness of the HOPG sheet and the structure parameters of the floating magnet are significantly related to the maximum moving space of the floating magnet ($P < 0.01$). In other words, changing the size of the lifting magnet does not cause a dramatic change in the maximum moving space $\sigma$ when the floating magnet and HOPG remain unchanged. The F value of the thickness of diamagnetic material, possessing a value of 22.95, establishes the thickness of HOPG as the most significant factor. From the analysis results, the influence factors of the maximum moving space are similar to those of the maximum gap analyzed by Simon M D et al. [18].

**Table 3.** Experimental layout using an L27 orthogonal array.

| Number | A | B | C | D | E | F | G | $\sigma$ (mm) | S/N |
|---|---|---|---|---|---|---|---|---|---|
| 1 | 1 | 1 | 1 | 1 | 1 | 1 | 1 | 1.54 | 3.7504 |
| 2 | 1 | 1 | 1 | 1 | 2 | 2 | 2 | 2.18 | 6.7691 |
| 3 | 1 | 1 | 1 | 1 | 3 | 3 | 3 | 2.30 | 7.2345 |
| 4 | 1 | 2 | 2 | 2 | 1 | 1 | 1 | 1.70 | 4.6089 |
| 5 | 1 | 2 | 2 | 2 | 2 | 2 | 2 | 2.26 | 7.0821 |
| 6 | 1 | 2 | 2 | 2 | 3 | 3 | 3 | 2.50 | 7.9588 |
| 7 | 1 | 3 | 3 | 3 | 1 | 1 | 1 | 1.78 | 5.0084 |
| 8 | 1 | 3 | 3 | 3 | 2 | 2 | 2 | 2.10 | 6.4443 |
| 9 | 1 | 3 | 3 | 3 | 3 | 3 | 3 | 2.14 | 6.6082 |
| 10 | 2 | 1 | 2 | 3 | 1 | 2 | 3 | 2.26 | 7.08217 |
| 11 | 2 | 1 | 2 | 3 | 2 | 3 | 1 | 1.44 | 3.16725 |
| 12 | 2 | 1 | 2 | 3 | 3 | 1 | 2 | 1.30 | 2.27887 |
| 13 | 2 | 2 | 3 | 1 | 1 | 2 | 3 | 2.04 | 6.1926 |
| 14 | 2 | 2 | 3 | 1 | 2 | 3 | 1 | 1.46 | 3.2870 |
| 15 | 2 | 2 | 3 | 1 | 3 | 1 | 2 | 1.64 | 4.2968 |
| 16 | 2 | 3 | 1 | 2 | 1 | 2 | 3 | 2.62 | 8.3660 |
| 17 | 2 | 3 | 1 | 2 | 2 | 3 | 1 | 2.40 | 7.6042 |
| 18 | 2 | 3 | 1 | 2 | 3 | 1 | 2 | 2.46 | 7.8187 |
| 19 | 3 | 1 | 3 | 2 | 1 | 3 | 2 | 1.50 | 3.5218 |
| 20 | 3 | 1 | 3 | 2 | 2 | 1 | 3 | 1.46 | 3.2870 |
| 21 | 3 | 1 | 3 | 2 | 3 | 2 | 1 | 0.88 | −1.1103 |
| 22 | 3 | 2 | 1 | 3 | 1 | 3 | 2 | 2.00 | 6.0206 |
| 23 | 3 | 2 | 1 | 3 | 2 | 1 | 3 | 2.00 | 6.0206 |
| 24 | 3 | 2 | 1 | 3 | 3 | 2 | 1 | 1.56 | 3.8624 |
| 25 | 3 | 3 | 2 | 1 | 1 | 3 | 2 | 2.20 | 6.8484 |
| 26 | 3 | 3 | 2 | 1 | 2 | 1 | 3 | 2.14 | 6.6082 |
| 27 | 3 | 3 | 2 | 1 | 3 | 2 | 1 | 1.68 | 4.5061 |

**Table 4.** Analysis of Variance for SN ratios for maximum moving space.

| Source | DF | Seq SS | Adj SS | Adj MS | F | P |
|---|---|---|---|---|---|---|
| Outer diameter of the floating magnet | 2 | 14.537 | 14.5375 | 7.2687 | 9.52 | 0.003 |
| Inner diameter of the floating magnet | 2 | 31.706 | 31.7057 | 15.8529 | 20.76 | 0.000 |
| Thickness of the floating magnet | 2 | 22.544 | 22.5441 | 11.272 | 14.76 | 0.001 |
| Outer diameter of the lifting magnet | 2 | 0.597 | 0.5971 | 0.2986 | 0.39 | 0.685 |
| Inner diameter of the lifting magnet | 2 | 4.106 | 4.1057 | 2.0528 | 2.69 | 0.108 |
| Thickness of the lifting magnet | 2 | 4.195 | 4.1951 | 2.0976 | 2.75 | 0.104 |
| Thickness of HOPG | 2 | 35.042 | 35.0425 | 17.5212 | 22.95 | 0.000 |
| Residual Error | 12 | 9.163 | 9.1632 | 0.7636 | | |
| Total | 26 | 121.891 | | | | |

### 4.2. Optimal Parameter Settings

Table 5 shows the mean value of S/N at each level corresponding to each factor. The effect value is the difference between the maximum and minimum S/N values of the factor at different levels. The importance of each factor affecting the maximum moving space $\sigma$ can be evaluated by the effect value, and the corresponding rank is also listed in Table 5. In addition, the effect of each factor on the S/N ratio is also illustrated in Figure 7. The optimal combination of parameters is A1B3C1D1E1F3G3.

**Table 5.** S/N value of each factor and level.

| Level | A | B | C | D | E | F | G |
|---|---|---|---|---|---|---|---|
| 1 | 6.163 | 3.998 | 6.383 | 5.499 | 5.711 | 4.853 | 3.854 |
| 2 | 5.566 | 5.481 | 5.571 | 5.460 | 5.586 | 5.466 | 5.676 |
| 3 | 4.396 | 6.646 | 4.171 | 5.166 | 4.828 | 5.806 | 6.595 |
| Effect | 1.767 | 2.648 | 2.212 | 0.333 | 0.883 | 0.953 | 2.742 |
| Rank | 4 | 2 | 3 | 7 | 6 | 5 | 1 |

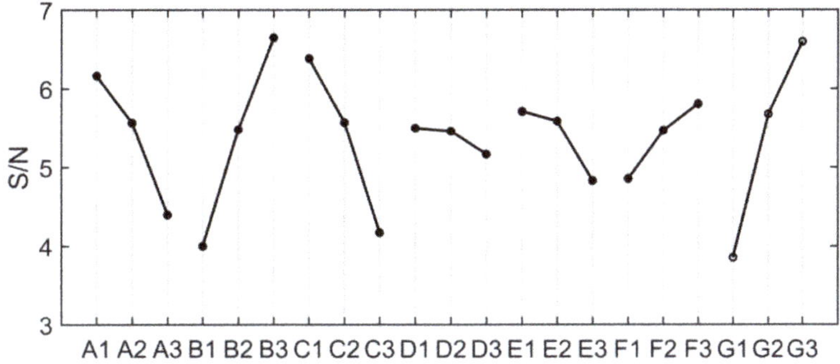

**Figure 7.** The larger the better S/N graph for the maximum moving space $\sigma$.

Besides, to verify the accuracy of the model, the 28th set of experiments was set according to the optimal structural parameters. MINITAB software predicted the maximum value of the moving space in the symmetric monostable structure to be 2.91 mm. The same optimization parameters are selected for the structural parameters to simulate and solve, and the maximum value of moving space $\sigma$ obtained is 2.78 mm. The error between the predicted value and the simulation value is only 4.467%, which is within the acceptable range.

## 5. Energy Harvesting Experiment

The diamagnetically stabilized levitation structure is adopted as the key component of an electromagnetic energy harvester with two coils fixed on the two pyrolytic graphite sheet. The whole device is packed within a shell. External excitation is applied to the shell of the harvester to make the internal floating magnet vibrate in the horizontal direction, and induced voltage is generated within the coils to realize the vibration energy harvesting. Based on parameters of the maximum space in the abovementioned symmetric monostable structure, the structure was decided in the experiments. Since magnetic flux gradient in the horizontal direction is significantly reduced when the floating magnet has a relatively large aperture, the inner diameter parameter of the levitation magnet is selected to be zero for energy harvesting experiments. The parameters of the experiment prototype are finally determined to be A1B1C1D2E1F3G3. The experiment setup is shown in Figure 8. For the specified dimension, the model predicts the maximum moving space to be 2.91 mm, and the measured one is 2.98 mm, with only 2.34% error. Through a vibration exciter (LT-50-ST250; ECON) connected with an acceleration sensor (EA-YD-188; ECON), vibration excitation is applied to the shell of the energy harvester. An oscilloscope (MOD3014) is used to measure the voltage signal generated at both ends of the coil.

According to the maximum moving space under the structural parameters, the selected coil parameters are 0.06 mm wire diameter, 5 mm inner coil diameter, 24.5 mm outer diameter, coil thickness about 0.72 mm, and the measured coil resistance is 724 $\Omega$. The coil is only arranged on the lower part of the upper HOPG. To indicate the moving space of the floating magnet after the coil is arranged on both sides, the paper with the same thickness as the upper coil is arranged on the lower side, as shown in Figure 9. The graphite plates with a low friction coefficient are used as the moving guide to ensure horizontal excitation without any additional load is exerted on the exciter. When the excitation peak value is set as 8 mm, the open-circuit voltage RMS at different frequencies is shown in Figure 10. When the excitation frequency is 2.6 Hz, the maximum voltage RMS reaches 250.69 mV and the power is 86.8 μW. The voltage waveform at the frequency is shown in Figure 11. If the coils are arranged on both sides of the floating magnet, the output voltage and power will be doubled. For the energy harvester, if the coil is directly connected to the Cockcroft–Walton cascade voltage doubler circuit [24] in the acquisition frequency range, the induced AC is rectified and boosted. In the experiment, an LED could be lit on after working for about 10 s, as shown in Figure 9. This excitation can be obtained by hand shaking.

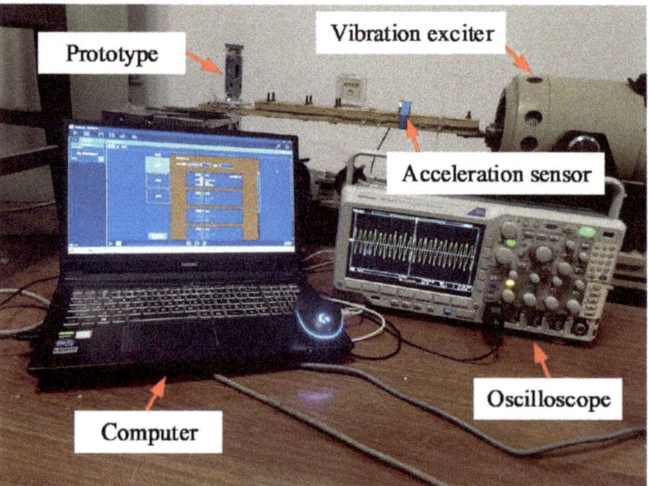


**Figure 8.** Experimental setup.

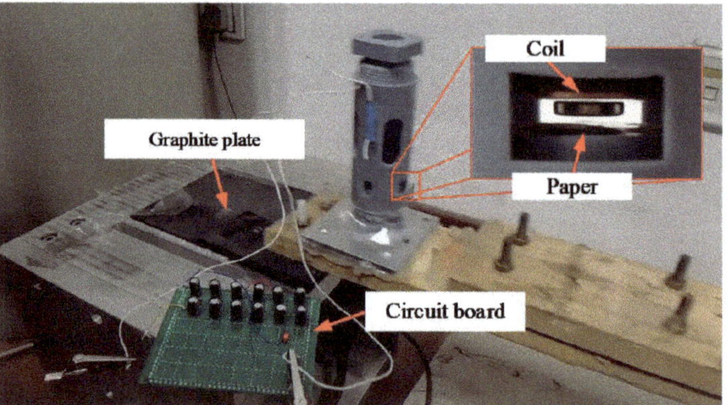


**Figure 9.** An LED illuminated by the energy harvester prototype.

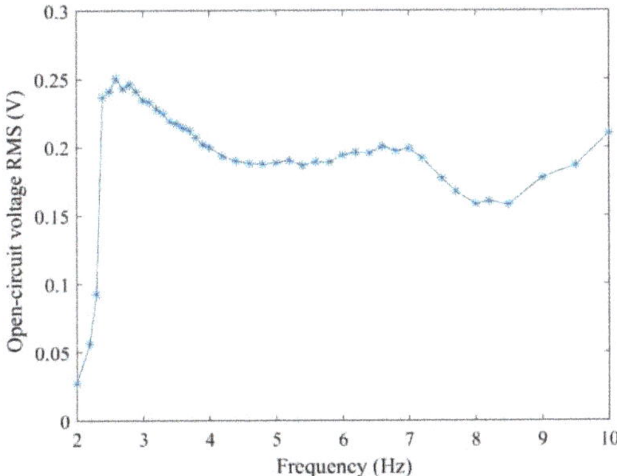

**Figure 10.** The RMS of voltage under different frequency excitation.

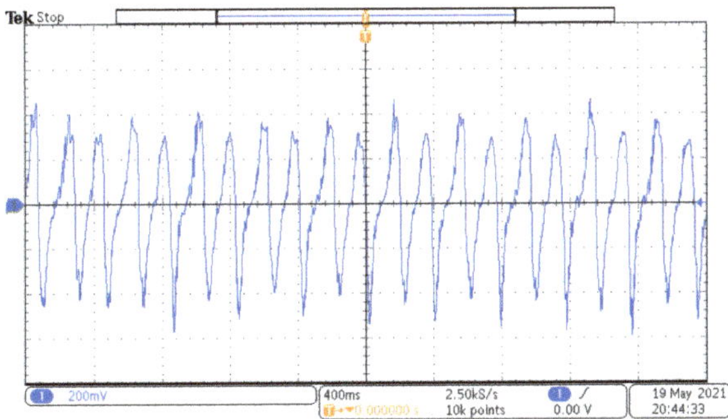

**Figure 11.** Voltage waveform at 2.6 Hz.

## 6. Conclusions

In this paper, the levitation characteristic of a diamagnetically stabilized levitation structure was investigated theoretically and experimentally. Three different stable levitation states were found by adjusting the gap between the two HOPG sheets, which includes the symmetric monostable levitation state, bistable levitation state, and asymmetric levitation state. The influence of structure parameters on the maximum moving space between two HOPG sheets in the symmetric monostable levitation structure was studied by the Taguchi method. According to the analysis, the maximum value of the moving space is mainly affected by the floating magnet and HOPG sheet. The thickness of HOPG sheets is the most important influence factor. Besides, the optimal combination of structure parameters is also determined. Through the prediction and verification of the optimized results, the accuracy of the model is proved. Using this analysis method, an optimal diamagnetically stabilized structure was built for actuating and sensing applications. A vibration energy harvester prototype was built based on the selected parameters. Experiments were carried out to verify the low-frequency performance of the energy harvester with maximum RMS voltage 250.69 mV and 86.8 µW power under 2.6 Hz excitation.

Compared with a model previously constructed by Ding et al. [13], the output power of the energy harvester was found to be increased by about 273.3%. The analysis and experimental results show that this method is effective for guiding the structural parameters. In future research work, the dynamic model of the energy harvester can be improved, coupled with the static model, and the output characteristics of the energy harvester can be optimized more directly.

**Author Contributions:** Conceptualization, Y.S.; methodology, S.C.; software, S.C. and Y.W.; validation, S.C. and Y.W.; formal analysis, S.C.; investigation, X.L.; resources, Y.S.; data curation, S.C.; writing—original draft preparation, S.C.; writing—review and editing, S.C.; visualization, X.L.; supervision, Y.S.; project administration, Y.S.; funding acquisition, Y.S. All authors have read and agreed to the published version of the manuscript.

**Funding:** This research was funded by the National Natural Science Foundation of China, grant number U1904169.

**Institutional Review Board Statement:** Not applicable.

**Informed Consent Statement:** Not applicable.

**Data Availability Statement:** Not applicable.

**Conflicts of Interest:** The authors declare no conflict of interest.

## References

1. Gao, Q.; Yan, H.; Zou, H. Magnetic levitation using diamagnetism: Mechanism, applications and prospects. *Sci. China Technol. Sci.* **2021**, *64*, 44–58. [CrossRef]
2. Braunbek, W. Freies schweben diamagnetischer körper im magnetfeld. *Z. Phys.* **1939**, *112*, 764–769. [CrossRef]
3. Simon, M.D.; Geim, A.K. Diamagnetic levitation: Flying frogs and floating magnets (invited). *J. Appl. Phys.* **2000**, *87*, 6200. [CrossRef]
4. Zhang, K.; Su, Y.; Ding, J.; Gong, Q.; Duan, Z. Design and Analysis of a Gas Flowmeter Using Diamagnetic Levitation. *IEEE Sens. J.* **2018**, *18*, 6978–6985. [CrossRef]
5. Clara, S.; Antlinger, H.; Abdallah, A.; Reichel, E.; Hilber, W.; Jakoby, B. An advanced viscosity and density sensor based on diamagnetically stabilized levitation. *Sens. Actuators A Phys.* **2016**, *248*, 46–53. [CrossRef]
6. Billot, M.; Piat, E.; Abadie, J.; Agnus, J.; Stempflé, P. External mechanical disturbances compensation with a passive differential measurement principle in nanoforce sensing using diamagnetic levitation. *Sens. Actuators A Phys.* **2016**, *238*, 266–275. [CrossRef]
7. Ho, J.N.; Wang, W. Electric generator using a triangular diamagnetic levitating rotor system. *Rev. Sci. Instrum.* **2009**, *80*, 24702. [CrossRef]
8. Palagummi, S.; Yuan, F.G. A bi-stable horizontal diamagnetic levitation based low frequency vibration energy harvester. *Sens. Actuators A Phys.* **2018**, *279*, 743–752. [CrossRef]
9. Gao, Q.; Zhang, W.; Zou, H.; Li, W.; Peng, Z.; Meng, G. Design and Analysis of a Bistable Vibration Energy Harvester Using Diamagnetic Levitation Mechanism. *IEEE Trans. Magn.* **2017**, *53*, 1–9. [CrossRef]
10. Palagummi, S.; Yuan, F.G. An optimal design of a mono-stable vertical diamagnetic levitation based electromagnetic vibration energy harvester. *J. Sound Vib.* **2015**, *342*, 330–345. [CrossRef]
11. Hilber, W.; Clara, S.; Jakoby, B. Sensing Physical Fluid Properties in Microcavities Utilizing Diamagnetic Levitation. *IEEE Trans. Magn.* **2015**, *51*, 1–4. [CrossRef]
12. Ye, Z.; Duan, Z.; Su, L.; Takahata, K.; Su, Y. Analysis of multi-direction low-frequency vibration energy harvester using diamagnetic levitation. *Int. J. Appl. Electrom.* **2015**, *47*, 847–860. [CrossRef]
13. Ding, J.; Gao, J.; Su, Y. A diamagnetically levitated vibration energy harvester for scavaging the horizontal vibration. *Mater. Res. Express* **2018**, *6*, 025506. [CrossRef]
14. Clara, S.; Antlinger, H.; Abdallah, A. A Viscosity and Density Sensor Based on Diamagnetically Stabilized Levitation. *IEEE Sens. J.* **2014**, *15*, 1937–1944. [CrossRef]
15. Liu, W.; Zhang, W.; Chen, W. Simulation analysis and experimental study of the diamagnetically levitated electrostatic micromotor. *J. Magn. Magn. Mater.* **2019**, *492*, 165634. [CrossRef]
16. Pujol-Vázquez, G.; Vargas, A.N.; Mobayen, S. Semi-Active Magnetic Levitation System for Education. *Appl. Sci.* **2021**, *11*, 5330. [CrossRef]
17. Chow, T.C.S.; Wong, P.L.; Liu, K.P. Shape Effect of Magnetic Source on Stabilizing Range of Vertical Diamagnetic Levitation. *IEEE Trans. Magn.* **2012**, *48*, 26–30. [CrossRef]
18. Simon, M.D.; Heflinger, L.O.; Geim, A.K. Diamagnetically stabilized magnet levitation. *Am. J. Phys.* **2001**, *69*, 702. [CrossRef]
19. Mia, M.; Dhar, N.R. Optimization of surface roughness and cutting temperature in high-pressure coolant-assisted hard turning using Taguchi method. *Int. J. Adv. Manuf. Technol.* **2017**, *88*, 739–753. [CrossRef]
20. Park, S.; Kim, H.; Kim, J. Taguchi Design of PZT-Based Piezoelectric Cantilever Beam with Maximum and Robust Voltage for Wide Frequency Range. *J. Electron. Mater.* **2019**, *48*, 6881–6889. [CrossRef]
21. Sheeraz, M.A.; Butt, Z.; Khan, A.M. Design and Optimization of Piezoelectric Transducer (PZT-5H Stack). *J. Electron. Mater.* **2019**, *48*, 6487–6502. [CrossRef]
22. Alrashdan, M.H.S.; Hamzah, A.A.; Majlis, B. Design and optimization of cantilever based piezoelectric micro power generator for cardiac pacemaker. *Microsyst. Technol.* **2015**, *21*, 1607–1617. [CrossRef]

23. Tikani, R.; Torfenezhad, L.; Mousavi, M.; Ziaei-Rad, S. Optimization of spiral-shaped piezoelectric energy harvester using Taguchi method. *J. Vib. Control* **2017**, *24*, 4484–4491. [CrossRef]
24. Abidin, N.A.K.Z.; Nayan, N.M.; Azizan, M.M.; Ali, A. Analysis of voltage multiplier circuit simulation for rain energy harvesting using circular piezoelectric. *Mech. Syst. Signal Process.* **2018**, *101*, 211–218. [CrossRef]

*micromachines*

**MDPI**

*Article*

# A Magnetic-Coupled Nonlinear Electromagnetic Generator with Both Wideband and High-Power Performance

**Manjuan Huang [1], Yunfei Li [2], Xiaowei Feng [1], Tianyi Tang [1], Huicong Liu [1,*], Tao Chen [1] and Lining Sun [1,2]**

[1] Jiangsu Provincial Key Laboratory of Advanced Robotics, School of Mechanical and Electric Engineering, Soochow University, Suzhou 215123, China; szdxhmj@126.com (M.H.); szdxfxw1014@163.com (X.F.); 20195229029@stu.suda.edu.cn (T.T.); chent@suda.edu.cn (T.C.); lnsun@hit.edu.cn (L.S.)

[2] Harbin Institute of Technology, School of Mechatronics Engineering, Harbin 215123, China; liyunfei3321@foxmail.com

* Correspondence: hcliu078@suda.edu.cn

**Abstract:** This paper proposed a high-performance magnetic-coupled nonlinear electromagnetic generator (MNL-EMG). A high-permeability iron core is incorporated to the coil. The strong coupling between the iron core and the vibrating magnets lead to significantly improved output power and a broadened operating bandwidth. The magnetic force of the iron core to the permanent magnets and the magnetic flux density inside the iron core are simulated, and the dimension parameters of the MNL-EMG are optimized. Under acceleration of 1.5 g, the MNL-EMG can maintain high output performance in a wide frequency range of 17~30 Hz, which is 4.3 times wider than that of linear electromagnetic generator (EMG) without an iron core. The maximum output power of MNL-EMG reaches 174 mW under the optimal load of 35 Ω, which is higher than those of most vibration generators with frequency less than 30 Hz. The maximum 360 parallel-connected LEDs were successfully lit by the prototype. Moreover, the prototype has an excellent charging performance such that a 1.2 V, 900 mAh Ni-MH battery was charged from 0.95 V to 0.98 V in 240 s. Both the simulation and experiments verify that the proposed bistable EMG device based on magnetic coupling has advantages of wide operating bandwidth and high output power, which could be sufficient to power micro electronic devices.

**Keywords:** vibration energy harvesting; electromagnetic generator (EMG); nonlinear; magnetic coupling; wideband; high performance

**Citation:** Huang, M.; Li, Y.; Feng, X.; Tang, T.; Liu, H.; Chen, T.; Sun, L. A Magnetic-Coupled Nonlinear Electromagnetic Generator with Both Wideband and High-Power Performance. *Micromachines* **2021**, *12*, 912. https://doi.org/10.3390/mi12080912

Academic Editor: Ju-Hyuck Lee

Received: 4 June 2021
Accepted: 20 July 2021
Published: 30 July 2021

**Publisher's Note:** MDPI stays neutral with regard to jurisdictional claims in published maps and institutional affiliations.



## 1. Introduction

With the maturity of low-power wireless transmission technology, various types of micro sensors, embedded systems, and wireless sensor networks have been developed rapidly [1]. The power consumption of micro electronic devices has been reduced to the order of microwatts. Due to the short service life, limited stored energy and environmental pollution problems, traditional chemical batteries are not suitable for the power supply of microelectronic devices with complex application environments [2–4].

Vibration energy harvesting technology converts vibration energy into electrical energy through electromechanical conversion mechanisms such as electromagnetic [5–8], piezoelectric [9–13], electrostatic [14–16], and triboelectric [17–20]. Among them, piezoelectric generators (PEGs) are especially attractive due to their simple construction, compact size, high power density and easy manufacturability, and PEGs are widely used in low-power electronic devices such as embedded systems and wireless sensing network nodes. Kim et al. [21] developed a flexible P(VDF-TrEE) film-based PEG on PDMS substrate with high power density, and the demonstrated output voltage, current, and power density were 5.8 V, 3.2 µA, and 6.62 mW/cm$^3$, respectively. Triboelectric nanogenerators (TENG), on the other hand, have attracted extensive attention and research because of its high output voltage, simple fabrication, low cost, and wide application of materials [22]. Han et al. [23]

presented an r-shaped hybrid piezo/triboelectric nanogenerator to enhance the output performance. With a 5 Hz periodic external force, the output voltage, current, and power density of the triboelectric part were 240 V, 27.2 $\mu$A, and 2.04 mW/cm$^3$. Ma et al. [24] designed and investigated an integrated electromagnetic-triboelectric-piezoelectric hybrid generator. The tested short-circuit currents of electromagnetic generator (EMG), TENG, and PEG are 21 mA, 4.1 $\mu$A, and 0.7 $\mu$A. Furthermore, the maximum instantaneous output powers of EMG, TENG, and PEG are 30.9 mW, 712.3 $\mu$W, and 6.37 $\mu$W, respectively, under the respective external load resistances of 200 $\Omega$, 70 M$\Omega$, and 50 M$\Omega$. Therefore, compared with PEG and TENG, EMG has the advantages of low impedance and high current output, and the output power can meet special requirements.

At present, most of the vibration generators are linear designs, which can achieve the maximum output power only when the resonance is consistent with the external vibration source [25]. Once the frequency of the vibration source deviates from the resonance of the generator, the output power will decrease rapidly, indicating a relatively narrow working bandwidth [26]. In general, the environmental vibration sources are random and cover a certain range, the working frequency of the vibration generators need to be broadened to achieve high output performance. To increase the output power and broaden the operating bandwidth of the vibration generators, various approaches, such as oscillator arrays [27], multi-modal oscillators [28,29], and active or passive frequency tuning technologies [30] have been proposed. In addition, researchers have exploited numerous approaches to introduce nonlinearity into vibration generators, including monostable Duffing oscillators [31,32], bistable oscillators [33–35], and frequency-up-conversion technologies [36,37] etc. The above mechanisms are commonly used in PEGs, EMGs, and TENGs. Stanton et al. [38] designed a bistable broadband PEG by adding a pair of repulsive magnets to a PZT cantilever. Bouhedma et al. [39] proposed a dual-frequency PEG with integrated magnets in a folded beam resonant structure. A bidirectional frequency tuning is achieved by adjusting the position of the magnet, which greatly broadens the operating bandwidth of the system. Magnetic spring is the most investigated structure among nonlinear EMGs. A traditional magnetic spring-based generator consists of two (top and bottom) fixed magnets with a third magnet levitated between them [40]. Mann et al. [41] proposed a Duffing electromagnet oscillator that uses magnetic levitation to realize resonance tuning and bandwidth broadening. Chen et al. [42] modelled and fabricated a multi-degree of freedom EMG using vertical linear springs and nonlinear magnetic springs based on magnetic levitation. In addition to magnetic spring structure, a double-clamped beam structure with strong natural nonlinearity is also an effective method to broaden the frequency bandwidth. Lu et al. [43] proposed a nonlinear electromagnetic vibration energy harvester with a monostable double-clamp beam, and the average power and frequency bandwidth of the harvester reached 1.78 mW and 11 Hz, respectively, at 1 g acceleration.

Although the bandwidth of EMGs can be broadened in several ways, most of them use air as the magnetic induction line transfer medium between the magnet and the coil, which leads to weak magnetic coupling because of the low magnetic permeability of air. When soft magnetic materials with high permeability are introduced into electromagnetic energy harvesting applications, it leads to significant enhancement in output voltage and power density compared to energy harvesters based on air-cored coils [44]. Soft magnetic materials with high magnetic permeability can be used either as flexible cantilevers in oscillating structures [45] or as fixed elements in coils [46]. However, soft magnetic materials as oscillating structures may induce high stiffness, and in this case low resonant frequencies cannot be achieved. In contrast, placing the soft magnetic material statically in the coil can significantly increase the power density by the strong magnetic coupling between the fixed high-permeability soft magnetic material and the oscillating permanent magnet pair. Sun et al. [44] designed an EMG using a pair of antiparallel permanent magnets as the vibrating part and a soft magnetic material with high permeability as the static part in the coil, which combine to form a closed magnetic circuit and make the coupling force between the permanent magnets and the soft magnetic material nonlinear by adjusting the spacing. This EMG can simultaneously increase the output power and operating bandwidth.

In this work, a simple but effective approach to realize the nonlinear EMG with both wide bandwidth and high power based on magnetic coupling between the magnets and iron core is presented. A high-permeability soft magnetic material is placed in the center of the coil as a permeable core, and the strong nonlinear magnetic coupling between the oscillating magnet and the core enhances the magnetic flux inside the coil, thus increasing the output power. Its design, mathematical model, and simulation methods can be beneficial to the research of other EMGs, which is the contribution of this paper. The structure design and working principle are described and the equivalent dynamical model is developed in Section 2. Simulation and parameter optimization are provided in Section 3. Experimental setup and detailed tests are described in Section 4, together with discussion of test results and comparison to other low-frequency vibration generators reported in recent literatures.

## 2. Device Design and Theoretical Analysis

*Structural Design and Working Principle*

The structure of the proposed magnetic-coupled nonlinear electromagnetic generator (MNL-EMG) is shown in Figure 1a. The MNL-EMG consists of a supporting base, a cantilever beam with two NdFeB magnets fixed at the free end, and a winding coil around a laminated iron core with high permeability. The material of the iron core is silicon lamination, which has 3% silicon content, which can increase the resistivity and maximum permeability of iron and reduce the core loss. The winding coil is wound with a diameter of 0.3 mm and a turn count of 500. The cantilever beam is made of phosphor bronze with good elasticity and non-magnetism. When the cantilever beam drives the end magnets to vibrate up and down, the iron core attracts the surrounding magnetic induction lines through the center of the coil. Therefore, the magnetic flux penetrating the coil changes, and an induced voltage will be generated across the coil. The magnetic force between the magnets and iron core varies nonlinearly with the displacement of the magnets. Hence the stiffness of the cantilever beam changes nonlinearly, resulting in the widening effect of operating bandwidth. Compared with the nonlinear EMGs relied on repulsion or attraction between magnets, this MNL-EMG based on the nonlinear attraction between the magnet and iron core has advantages of increasing the electromagnetic output power using a high-permeability iron core. Figure 1b,c shows the distribution of magnetic induction lines around the magnets and iron core when the magnets are above or below the iron core. The magnetic induction lines start from the N pole of the magnets, flow into the iron core, and then flow back to the S pole of the magnets. The iron core with high permeability can concentrate the magnetic induction lines inside the coil, enable the direction of magnetic induction lines perpendicular to the coil plane, greatly increases the magnetic flux, therefore increasing the induced voltage of the coil. The schematic diagram of the MNL-EMG is shown in Figure 1d, and the geometric parameters are listed in Table 1.

An equivalent mass-spring-damper model of the MNL-EMG is developed as shown in Figure 2a. The corresponding dynamic equation can be expressed as:

$$m_{eq}Z(t) + c_{eq}Z(t) + k_{eq}Z(t) = -m_{eq}u(t) - F_z(Z) - F_e(Z),\qquad(1)$$

where $m_{eq}$, $c_{eq}$, and $k_{eq}$ represent the equivalent mass, equivalent damping, and equivalent stiffness of the cantilever beam, respectively; $Z$ is the vertical displacement of the magnets at the free end of the cantilever beam; $u(t)$ is the displacement of the external sinusoidal excitation; $F_z$ is the vertical component of the magnetic force $F_m$ between the magnets and iron core; and $F_e$ is the electromagnetic damping force between the magnets and coil. It is assumed that the displacement function $u(t)$ of the sinusoidal excitation and the nonlinear magnetic force function $F_z(Z)$ are

$$u(t) = \frac{a}{\omega^2}\sin(\omega t),\qquad(2)$$

and

$$F_z(Z) = p_0 + p_1 Z + \cdots\cdots + p_n Z^n, \tag{3}$$

where $a$ and $\omega$ are respectively the acceleration and angular frequency of the sinusoidal excitation; $p_0, p_1 \cdots\cdots p_n$ are the coefficients of the polynomial function $F_z(Z)$.

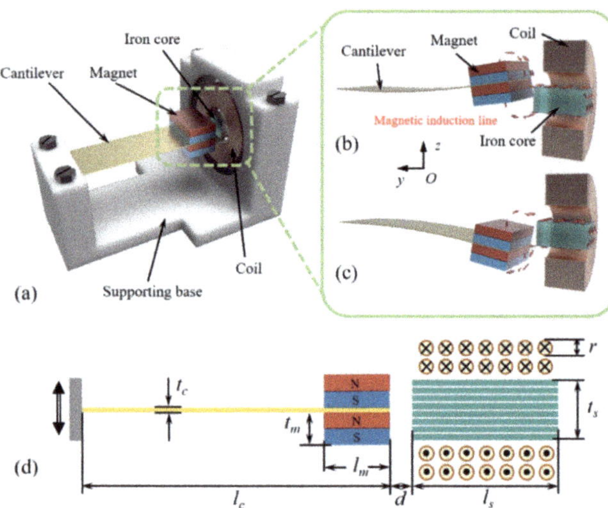


**Figure 1.** (a) Structure of the magnetic-coupled nonlinear electromagnetic generator (MNL-EMG). Distribution of magnetic induction lines around the magnets and iron core as the magnets are (b) above or (c) below the iron core. (d) Schematic diagram of the MNL-EMG.

**Table 1.** Geometric dimensions of the MNL-EMG.

| Parameter | Value |
|---|---|
| Length of the cantilever beam $l_c$ | 40 mm |
| Width of the cantilever beam $W_c$ | 13 mm |
| Thickness of the cantilever beam $t_c$ | 0.3 mm |
| Length of the magnet $l_m$ | 10 mm |
| Width of the magnet $W_m$ | 15 mm |
| Thickness of the magnet $t_m$ | 5 mm |
| Length of the iron core $l_s$ | 15 mm |
| Width of the iron core $W_s$ | 7 mm |
| Thickness of the iron core $t_s$ | 7 mm |
| Distance between the magnet and iron core $d$ | 2 mm |
| Diameter of the copper wire $r$ | 0.3 mm |
| Diameter of the coil $R$ | 30 mm |
| Turns of the coil $N$ | 500 |

The power generated by the electromagnetic damping force $F_e$ is equal to the product of the electromagnetic damping force and the relative movement speed, then the power is mainly consumed on the external load $R_{load}$ and the internal resistance $R_{coil}$ of the coil. Therefore, the instantaneous power $P_e$ can be expressed as

$$P_e = F_e \frac{dZ}{dt} = \frac{E^2}{R_{coil} + R_{load}}, \tag{4}$$

where $E$ is the induced voltage of the coil, $dZ/dt$ is the moving speed of the magnet. According to the law of electromagnetic induction, the induced voltage generated by the change of the magnetic flux in the closed coil can be expressed as [6]

$$E = -\frac{d\Psi}{dt} = -N\frac{d\Phi}{dt},\tag{5}$$

where $\Psi$ is the total magnetic flux, $N$ is the turns of the coil, $\Phi$ is the magnetic flux passing through the single-turn coil, and $\Phi$ can be expressed as

$$\Phi = BS\cos\theta,\tag{6}$$

where $B$ is the magnetic flux density of each turn of the coil, $S$ is the area enclosed by the coil, and $\theta$ is the angle between the normal vector of the coil plane and the direction of the magnetic field. It can be seen from Figure 1 that the angle between the normal vector of the coil plane and the direction of the magnetic field inside the iron core is equal to 0. By substituting Equations (5) and (6) into Equation (4), the electromagnetic damping force can be further expressed as

$$F_e = NS\frac{dB}{dZ}\frac{E}{R_{coil} + R_{load}},\tag{7}$$

where $dB/dZ$ is the gradient of the magnetic flux density along $z$-axis.

To calculate the electromechanical dynamics in a couple manner, this whole model with all above theories is established in MATLAB/Simulink, as shown in Figure 2b. The geometric dimensions used in the model are listed in Table 1.

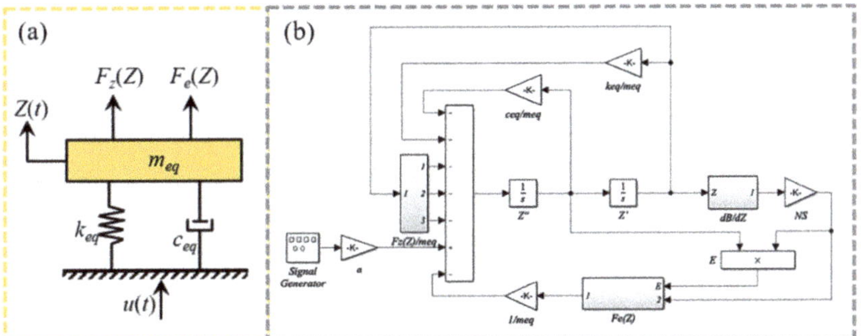


**Figure 2.** (**a**) Equivalent model and (**b**) MATLAB/Simulink diagram of the MNL-EMG.

## 3. Simulation Analysis and Optimization

*Analysis of Nonlinear Dynamics*

Finite element analysis software COMSOL Multiphysics is used to simulate the magnetic field intensity and distribution at different displacements, and the simulation model is shown in Figure 3. The simulation results are shown in Figure 4. Placing an iron core near the magnets can guide the magnetic induction lines flowing from N pole of the magnets into the iron core, and then return to S pole after passing through the iron core. In addition to magnetic flux density inside the iron core, the magnetic force between the magnets and iron core can be also calculated by the magnetic field simulation. Using the curve fitting module in MATLAB, the polynomial function expression between the vertical magnetic force $F_z$ and the displacement $Z$ is fitted as Equation (3), the polynomial function expression between the horizontal magnetic flux density $B_y$ and the displacement $Z$ are fitted as

$$B_y(Z) = q_0 + q_1 Z + \cdots\cdots + q_m Z^m,\tag{8}$$

where $q_0, q_1 \cdots \cdots q_m$ are the coefficients of the polynomial $B_y(Z)$. The $dB_y/dZ$ can be used to calculate the induced voltage of the coil.

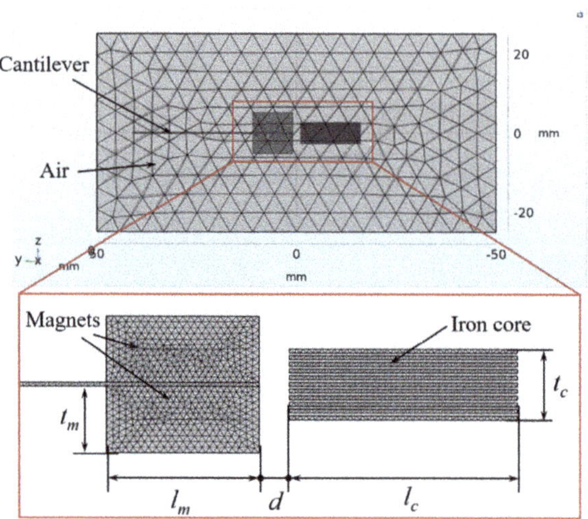


**Figure 3.** COMSOL magnetic field simulation model diagram.

Figure 5 shows the comparison of the simulated scatter points and the fitted curves at different displacements Z from −20 to 20 mm when the initial gap distance $d$ between the magnets and iron core is 2 mm. Within a certain error range, the variation trend and numerical value of the simulated scatter points and the fitted curves are basically coincided. Figure 5a shows the fitted curve of the nonlinear magnetic force $F_z$. When the value of $F_z$ is positive, it means that the magnetic force of the iron core to magnets is along the positive direction of z-axis. Conversely, when the value of $F_z$ is negative, the magnetic force of the iron core to magnet is along the negative direction of z-axis. As the displacement decreases from 20 mm to 0, the magnets are initially attracted by the iron core, and the attractive force gradually increases to the maximum value at displacement of 12 mm (point A). Then the attractive force drops to zero at displacement of 8 mm (point B). Next, the magnets are repelled by the iron core, and the repulsive force reaches the maximum value at 4 mm (point C), then decreases to zero at displacement of 0 (point D). Since the curve of $F_z$ is symmetrical about the origin of the coordinate axis, the magnets are repelled first and then attracted when the displacement decreases from 0 to −20 mm. The maximum repulsive force, the maximum attractive force, and the magnetic force equal to zero appear at −4 (point E), −12 (point F), and −8 mm (point G), respectively. After the magnetic force reaches the peak point G, the distance between the magnet and the iron core increases as the displacement increases, and the magnetic field strength around the magnet decreases, so the magnetic force decreases.

Correspondingly, the distribution of the magnetic field intensity and magnetic induction lines at point A to G is shown in Figure 4a–g, respectively, and the fitted curve of the magnetic flux density $B_y$ inside the iron core is depicted in Figure 5b. As shown in Figure 4a,g, the direction of magnetic induction lines inside the iron core at points A and G towards the positive and negative directions of y-axis, respectively. The iron core sheet closest to the magnet has the highest magnetic flux density, while the other sheets have very low magnetic flux density. The magnetic induction lines distribution at points B and F are shown in Figure 4b,f. It is seen that the magnetic induction lines introduced into the iron core are unidirectional and the number is the largest, so magnitude of $B_y$ at these two points is the highest, as shown in Figure 5b. At points C (Figure 4c) and E (Figure 4e), since the iron core is close to the N and S poles of the magnets, the magnetic induction lines inside the iron core are bidirectional, resulting in a decrease of $B_y$. The magnetic induction lines start from the N pole into the iron core, then flow out from the iron core to the S pole. Figure 4d is the magnetic flux density distribution at point D. Here the N and S poles of the magnets are located on both sides of the iron core, the distance between the two magnetic poles and the center of the iron core is equal, so the magnetic induction lines flowing in and out of the iron core are equal. Thus, the magnetic flux density $B_y$ decreases to zero at point D.

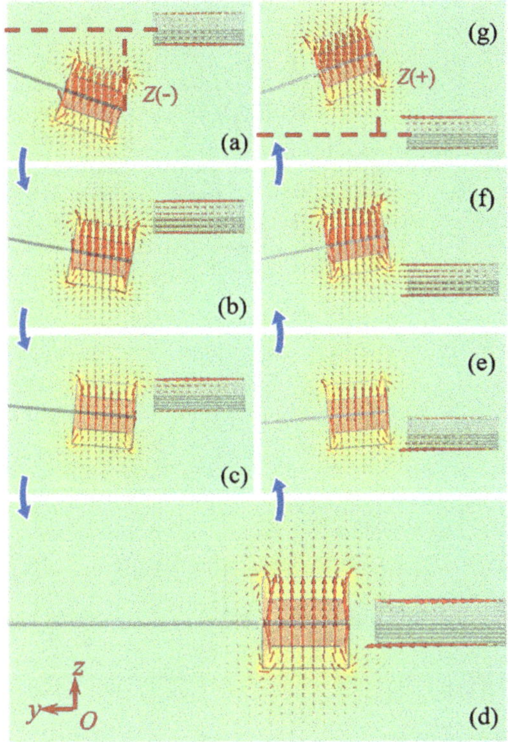


**Figure 4.** The magnetic field intensity and distribution at different displacements: (**a**) $Z = 12$ mm (point A); (**b**) $Z = 8$ mm (point B); (**c**) $Z = 4$ mm (point C); (**d**) $Z = 0$ (point D); (**e**) $Z = -4$ mm (point E); (**f**) $Z = -8$ mm (point F); (**g**) $Z = -12$ mm (point G).

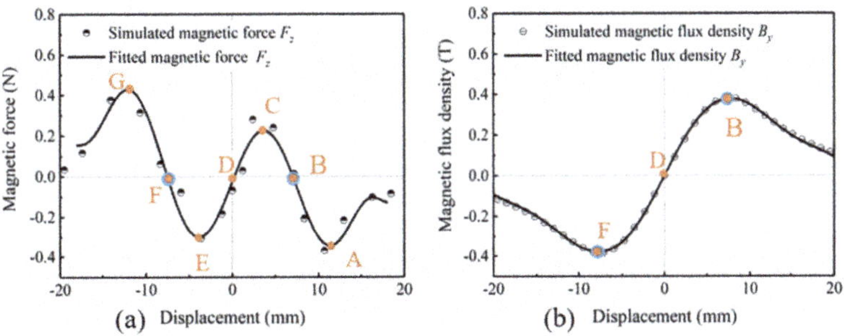


**Figure 5.** Fitted curves of (**a**) magnetic force $F_z$ and (**b**) magnetic flux density $B_y$ at distance of 2 mm.

Based on the COMSOL magnetic field simulation and MATLAB curve fitting, the functional expression of the nonlinear magnetic force $F_z(Z)$ and the gradient $dB_y/dZ$ of the magnetic flux density inside the iron core is obtained. According to the mathematical model, the voltage induced by the coil of the MNL-EMG under sinusoidal excitation is simulated. Since the gap distance between the magnets and iron core strongly affects the nonlinear magnetic force, the nonlinear response of the cantilever beam and the operating bandwidth of the MNL-EMG are simulated against different gap distances. Figure 6a–c depict the open-circuit voltage of the MNL-EMG with cantilever beam thickness of 0.3 and 0.4 mm against the frequency up-sweep as the gap distance $d$ varying from 1, 2 to 3 mm, respectively, under acceleration of 1.5 g. It is seen that the MNL-EMG with beam thickness of 0.3 mm has higher voltage output and lower resonant frequency than that with beam thickness of 0.4 mm. The maximum output voltage occurs at gap distance of 2 mm and resonant frequency of 26 Hz. This is because the closer the magnets to the iron core, the greater the magnetic attraction along the $y$-axis direction. Once the gap distance is less than 1 mm, the magnet will be firmly attracted by the iron core, resulting in extremely low output voltage. As the gap distance increases, the attractive force along the $y$-axis direction decreases and the output voltage increases. However, the nonlinear magnetic force along the $z$-axis decreases with the increase of the distance. The nonlinearity of the voltage would be gradually weakened as the gap distance reaches 3 mm. The simulation results show that the MNL-EMG with a gap distance of 2 mm combines high output voltage and wide operating bandwidth. Figure 6d shows the phase trajectory and displacement of the MNL-EMG with gap distance of 2 mm when the excitation amplitude $a$ is 1.5 g and the excitation frequency $f$ is 26 Hz. The MB-EMG performs a large-amplitude periodic oscillation.

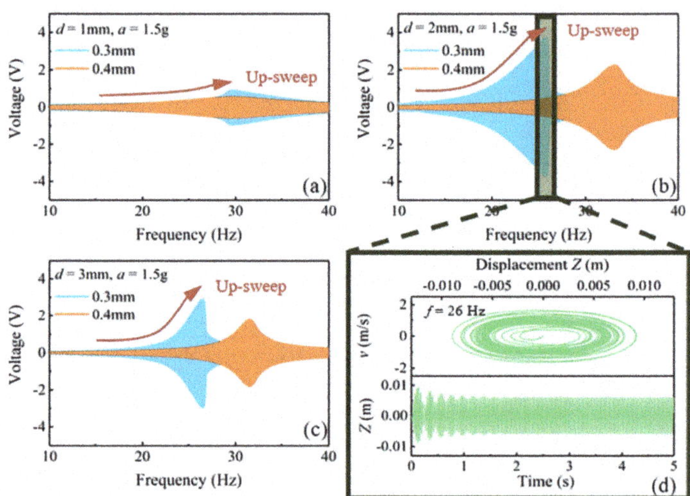


**Figure 6.** The simulated open-circuit voltage of the MNL-EMG with beam thickness of 0.3 mm and 0.4 mm, under gap distance of (**a**) 1 mm, (**b**) 2 mm, and (**c**) 3 mm. (**d**) The phase trajectory and displacement of the MNL-EMG with gap distance of 2 mm when the excitation acceleration $a$ is 1.5 g and the excitation frequency $f$ is 26 Hz.

## 4. Experimental Results and Discussion

### 4.1. Experimental Setup

The output performance of the MNL-EMG was tested by the vibration control and testing system as shown in Figure 7. The system setup consists of a computer, a vibration exciter (TIRA TV 511110, Germany), a power amplifier (BAA 120), a signal generator (Vipilot 4 channels) and an accelerometer (DYTRAN model 3035BG, USA). The MNL-EMG prototype is fixed on the vibration exciter, and the sinusoidal excitation of the vibration exciter is generated by the signal generator through the power amplifier. The accelerometer is installed on the exciter to monitor the excitation acceleration and realize the closed-loop feedback control of the vibration control system. The frequency and acceleration of the sinusoidal excitation can be adjusted by the control software in the computer. The voltage output of the MNL-EMG in frequency and time domains can be accurately measured and recorded by the dynamic signal analysis software (m+p international VibRunner).

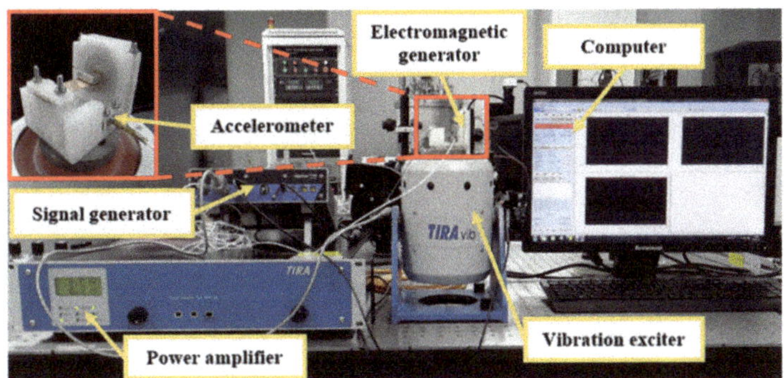


**Figure 7.** The experimental setup and the prototype.

### 4.2. Results and Discussion

First, the EMG was subjected to a frequency upward sweep test using the vibration control and test system in the range of 10~40 Hz. As shown in Figure 8, the output performance of the EMG with and without iron core was compared under vibration acceleration of 1.5 g and gap distance of 2 mm. The resonant frequency of the EMG without iron core is 27 Hz, and the maximum output voltage at 27 Hz is 1.3 V. It is seen clearly that the voltage output of the MNL-EMG in frequency domain exhibits nonlinear characteristics, as shown in Figure 8a. The operating bandwidth of the MNL-EMG is 13 Hz (from 17 to 30 Hz), which is 4.3 times wider than that of the EMG without iron core, which is only 3 Hz (from 25.6 to 28.6 Hz). In addition, the high magnetic permeability of the iron core resulting in the increase of magnetic flux density inside the coil, so the output voltage of the MNL-EMG increases significantly. The maximum open-circuit voltage of the MNL-EMG (3.8 V at 29 Hz) is about 3 times higher than that of the EMG without iron core (1.3 V at 27 Hz). It is verified that the use of iron core with high magnetic permeability to the coil can broaden the operating bandwidth and improve the output performance effectively. The time domain diagrams of the open-circuit voltage of the EMG with iron core and without iron core are depicted in Figure 8b.

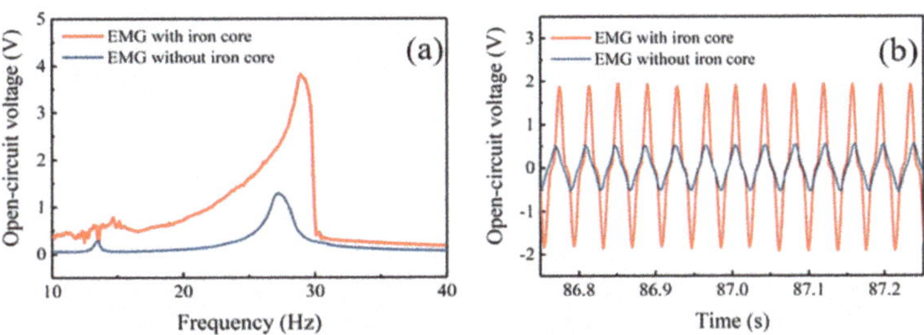


**Figure 8.** (**a**) Frequency domain and (**b**) time domain diagrams of the open-circuit voltage of the EMG with iron core and without iron core.

The output performance of the MNL-EMG is strongly affected by the gap distance $d$ between the magnet and iron core, as well as the acceleration of the excitation. Figure 9 shows the measured open-circuit voltage of the MNL-EMG against operating frequency at various acceleration levels of 0.25, 0.5, 1, and 1.5 g, under the gap distance of 1, 2, and 3 mm, respectively. It is seen the open-circuit voltage outputs of the MNL-EMG are quite low at the gap distance of 1 mm. This is because the magnetic attraction along $y$-axis is very strong, the deflection of the cantilever beam is strongly restricted by the magnetic attraction force. In contrast, under gap distance of 2 mm, the open-circuit voltage of the MNL-EMG is 2.39, 3.33 and 3.82 V at acceleration of 0.5, 1.0 and 1.5 g, respectively, which are higher than those of the MNL-EMG under gap distance of 3 mm. The maximum output voltage is 3.82 V at excitation frequency of 30 Hz. It should be noted that the open-circuit voltage of the MNL-EMG at acceleration of 0.25 g and under gap distance of 2 mm is 0.37 V, which is much lower than that of the MNL-EMG under gap distance of 3 mm. It can be inferred that greater acceleration is required to enhance the nonlinear response of the cantilever, therefore increasing the output voltage and broadening the bandwidth. Figure 10 shows the comparison of the simulated and tested waveform diagrams of the open-circuit voltage against frequency up-sweep at vibration acceleration of 1.5 g and under gap distance of 2 mm. The waveforms of the open-circuit voltage obtained by simulation and experiment are consistent, indicating that the established system dynamic model is suitable for this MNL-EMG device.

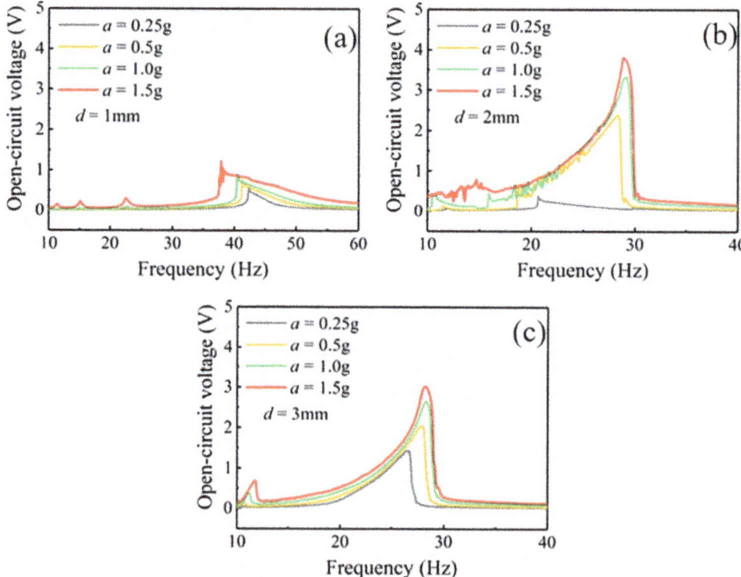


**Figure 9.** The open-circuit voltage outputs of the MNL-EMG at different accelerations under the distance $d$ of (**a**) 1 mm, (**b**) 2 mm and (**c**) 3 mm, respectively.

Figure 11a shows the peak load voltage and maximum output power of the MNL-EMG versus load resistance at acceleration of 1.5 g and frequency of 30 Hz. The MNL-EMG can be regarded as a voltage source with an open-circuit voltage of $E$ and an internal resistance of $r$ equal to the resistance of the coil. With the increasing of load resistance, the load voltage continues to increase, while the corresponding power reaches to a maximum value of 174 mW at the optimal load resistance of 35 $\Omega$. The maximum output power of MNL-EMG is 10.8 times higher than that of the EMG without iron core. Figure 11b demonstrated that the MNL-EMG can light up more than 360 parallel-connected LEDs.

Moreover, to examine the charging performance of the MNL-EMG, the output of the MNL-EMG is connected to a rectifier to convert the AC input signal into a DC signal. Then the HT7335 voltage regulator chip converts the rectified DC voltage into a stable DC voltage to charge the battery, as shown in Figure 12a. The rechargeable battery is a Ni-MH battery with 1.2 V nominal voltage and 900 mAh nominal capacity. Figure 12b presents the voltage of the battery during the experiment. The battery voltage was 0.95 V at the beginning. After the charging process of 240 s, the stable voltage of the rechargeable battery increased to 0.98 V. In real applications, the MNL-EMG is designed to harvest the vibration energy generated by these mechanical devices, such as vibrating screen, coal mining machine, pump, gear box, etc., to power the sensors and eventually build a self-powered wireless sensing network.

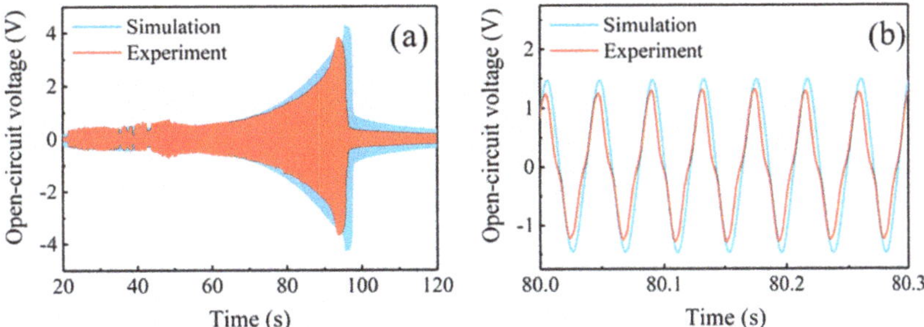

**Figure 10.** (**a**,**b**) The tested and simulated time domain waveform diagrams of the open-circuit voltage of the MNL-EMG at acceleration of 1.5 g, under distance of 2 mm.

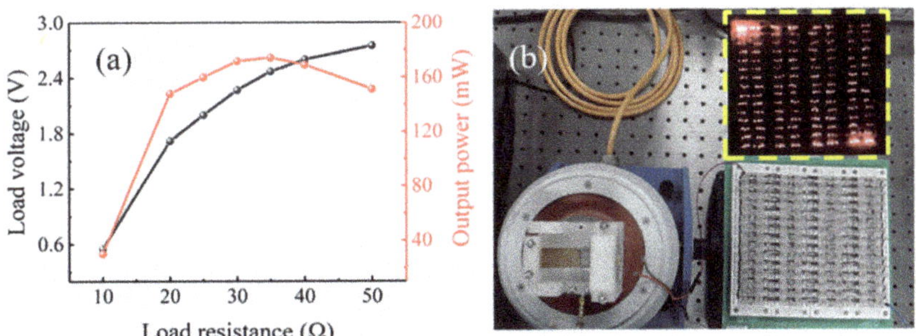

**Figure 11.** (**a**) The load voltage and maximum output power of the MNL-EMG versus load resistance. (**b**) Photograph of LEDs that are lighted up by MNL-EMG.

Performance comparison between the introduced MNL-EMG and recent published low-frequency vibration generators in the literature is given in Table 2. It is worth pointing out that a fair comparison may need to re-design the energy harvester for the specific working condition, which is quite difficult. The acceleration, operating frequency range, and power density somehow generalized working condition and performance parameter. As the data shown in Table 2, previous works investigated different transaction mechanisms for the low-frequency vibration. The MNL-EMG in this work can generate a high power output and a wide bandwidth at relatively large acceleration, which indicates its superior performance over other vibration generators in the large-amplitude vibration environment. In general, the introduced MNL-EMG device has achieved high output power, wide working bandwidth, and low working frequency, simultaneously. It can easily capture the low-frequency environmental vibration energy, and the output power is sufficient to power micro electronic devices. In the subsequent research, we will continue the integration of the power management circuit for the MNL-EMG and build a complete power supply system, which would have broad application prospects.

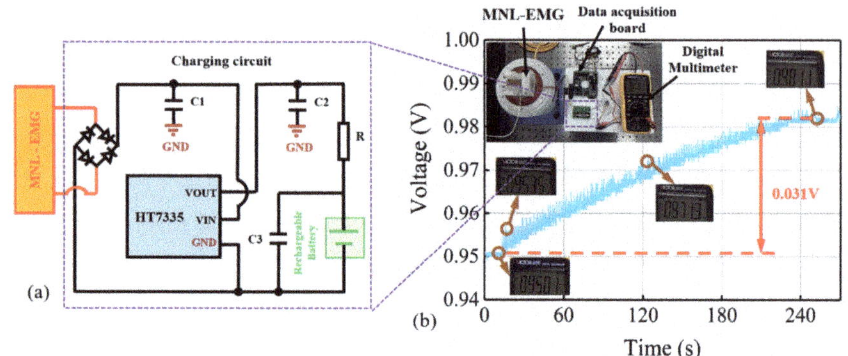

**Figure 12.** (**a**) Experimental circuit and (**b**) battery charging experiment.

**Table 2.** Performance comparison between the MNL-EMG and low-frequency vibration generators reported in the literature.

| Reference | Transducer | Acceleration (g) | Frequency (Hz) | Bandwidth (Hz) | Power (mW) |
| --- | --- | --- | --- | --- | --- |
| Ren [47] | PE | 0.2 | 12~14.5 | 2.5 | 1.40 |
| Shen [6] | EM | 3.25 | 3~6 | 3 | 23.2 |
| Yan [7] | EM | 0.6 | 5~15 | 10 | 28 |
| Aldawood [40] | EM | 0.4 | 7~12 | 5 | 80 |
| Sun [44] | EM | 0.5 | 57.4~64.7 | 7.3 | 0.003 |
| Haroun [48] | EM | 1.26 | 3.33 | – | 0.083 |
| Gu [49] | EM | 1.0 | 5~27 | 22 | 7.65 |
| Toyabur [27] | EM and PE | 0.4 | 12~22 | 10 | 0.49 |
| Fan [50] | EM and PE | 1.5 | 6~8.5 | 2.5 | 1.42 |
| Salauddin [51] | EM and TE | 0.5 | 5 | – | 11.75 |
| Askari [52] | EM and TE | 0.57 | 45.45~55.45 | 10 | 74 |
| This work | EM | 1.5 | 17~30 | 13 | 174 |

$g$ = 9.8 m/s$^2$, PE = piezoelectric, EM = electromagnetic, TE = triboelectric.

## 5. Conclusions

This paper presents a MNL-EMG with wide bandwidth and high performance. In this MNL-EMG, advantages of several technologies are combined to enhance the output performance: (1) the nonlinear attraction between magnets and iron core enable the cantilever beam to operate in a high-energy orbit to generate wider operating bandwidth and higher output power; (2) iron core with high-permeability guides the magnetic induction lines into the coil, therefore greatly increased the electromagnetic output power while performing nonlinear motion. To reveal the mechanism of nonlinear effect, the magnetic force of the iron core to the magnets and the magnetic flux density inside the iron core at different displacements are simulated and analyzed. Based on the dynamic model, the output voltage of the MNL-EMG is simulated, and the dimension parameters are optimized. Then a series of experiments are conducted to test the output performance of the prototype. Under an acceleration of 1.5 g, the MNL-EMG can maintain high output power in the ultra-wide frequency range of 17~30 Hz, and the operating bandwidth reaches 13 Hz, which is 4.3 times wider than that of linear EMG without an iron core. The maximum output power of the MNL-EMG reaches 174 mW under the optimal load resistance of 35 Ω, and the prototype can light up 360 LEDs. Moreover, the prototype can charge a Ni-MH rechargeable battery (1.2 V, 900 mAh) from 0.95 V to 0.98 V within 240 s. It is proved that the proposed MNL-EMG has advantages of wide operating bandwidth and extremely high output power. This research can be of great significance for further exploration of nonlinear vibration energy harvester.

**Author Contributions:** Conceptualization, M.H. and H.L.; methodology, M.H. and X.F.; software, M.H. and Y.L.; validation, M.H. and T.T.; formal analysis, M.H. and T.C.; writing—original draft preparation, M.H. and X.F.; writing—review and editing, M.H. and H.L.; project administration, H.L. and T.C.; funding acquisition, H.L. and L.S. All authors have read and agreed to the published version of the manuscript.

**Funding:** This research was funded by the National Key Research and Development Program of China (Grant No. 2019YFB2004800) and the National Natural Science Foundation of China (Grant No. 51875377).

**Acknowledgments:** The authors gratefully acknowledge the support from the National Key Research and Development Program of China (Grant No. 2019YFB2004800) and the National Natural Science Foundation of China (Grant No. 51875377).

**Conflicts of Interest:** The authors declare no conflict of interest.

# References

1. Liu, H.; Zhong, J.; Lee, C.; Lee, S.-W.; Lin, L. A comprehensive review on piezoelectric energy harvesting technology: Materials, mechanisms, and applications. *Appl. Phys. Rev.* **2018**, *5*, 041306. [CrossRef]
2. Zou, H.X.; Zhao, L.C.; Gao, Q.H.; Zuo, L.; Liu, F.R.; Tan, T.; Wei, K.X.; Zhang, W.M. Mechanical modulations for enhancing energy harvesting: Principles, methods and applications. *Appl. Energy* **2019**, *255*, 113871. [CrossRef]
3. Hou, C.; Chen, T.; Li, Y.; Huang, M.; Shi, Q.; Liu, H.; Sun, L.; Lee, C. A rotational pendulum based electromagnetic/triboelectric hybrid-generator for ultra-low-frequency vibrations aiming at human motion and blue energy applications. *Nano Energy* **2019**, *63*, 103871. [CrossRef]
4. Yang, Z.; Zhu, Y.; Zu, J. Theoretical and experimental investigation of a nonlinear compressive-mode energy harvester with high power output under weak excitations. *Smart Mater. Struct.* **2015**, *24*, 025028. [CrossRef]
5. Liu, H.; Hou, C.; Lin, J.; Li, Y.; Shi, Q.; Chen, T.; Sun, L.; Lee, C. A non-resonant rotational electromagnetic energy harvester for low-frequency and irregular human motion. *Appl. Phys. Lett.* **2018**, *113*, 203901. [CrossRef]
6. Shen, Y.; Lu, K. Scavenging power from ultra-low frequency and large amplitude vibration source through a new non-resonant electromagnetic energy harvester. *Energy Convers. Manag.* **2020**, *222*, 113233. [CrossRef]
7. Yan, B.; Yu, N.; Zhang, L.; Ma, H.; Wu, C.; Wang, K.; Zhou, S. Scavenging vibrational energy with a novel bistable electromagnetic energy harvester. *Smart Mater. Struct.* **2020**, *29*, 025022. [CrossRef]
8. Li, Y.; Guo, Q.; Huang, M.; Ma, X.; Chen, Z.; Liu, H.; Sun, L. Study of an Electromagnetic Ocean Wave Energy Harvester Driven by an Efficient Swing Body Toward the Self-Powered Ocean Buoy Application. *IEEE Access* **2019**, *7*, 129758–129769. [CrossRef]
9. Huang, M.; Hou, C.; Li, Y.; Liu, H.; Wang, F.; Chen, T.; Yang, Z.; Tang, G.; Sun, L. A Low-Frequency MEMS Piezoelectric Energy Harvesting System Based on Frequency Up-Conversion Mechanism. *Micromachines* **2019**, *10*, 639. [CrossRef] [PubMed]
10. Tang, L.; Yang, Y. A nonlinear piezoelectric energy harvester with magnetic oscillator. *Appl. Phys. Lett.* **2012**, *101*, 094102. [CrossRef]
11. Yang, B.; Yi, Z.; Tang, G.; Liu, J. A gullwing-structured piezoelectric rotational energy harvester for low frequency energy scavenging. *Appl. Phys. Lett.* **2019**, *115*, 063901. [CrossRef]
12. Cao, J.; Lin, J.; Inman, D.J.; Zhou, S. Nonlinear Dynamic Characteristics of Variable Inclination Magnetically Coupled Piezoelectric Energy Harvesters. *J. Vib. Acoust.* **2015**, *137*, 021015. [CrossRef]
13. Machado, L.Q.; Yurchenko, D.; Wang, J.; Clementi, G.; Margueron, S.; Bartasyte, A. Multi-dimensional constrained energy optimization of a piezoelectric harvester for E-gadgets. *iScience* **2021**, *24*, 102749. [CrossRef]
14. Tao, K.; Lye, S.W.; Miao, J.; Tang, L.; Hu, X. Out-of-plane electret-based MEMS energy harvester with the combined nonlinear effect from electrostatic force and a mechanical elastic stopper. *J. Micromech. Microeng.* **2015**, *25*, 104014. [CrossRef]
15. Vysotskyi, B.; Aubry, D.; Gaucher, P.; Le Roux, X.; Parrain, F.; Lefeuvre, E. Nonlinear electrostatic energy harvester using compensational springs in gravity field. *J. Micromech. Microeng.* **2018**, *28*, 074004. [CrossRef]
16. Zhang, Y.; Wang, T.; Luo, A.; Hu, Y.; Li, X.; Wang, F. Micro electrostatic energy harvester with both broad bandwidth and high normalized power density. *Appl. Energy* **2018**, *212*, 362–371. [CrossRef]
17. Shi, Q.; Wang, H.; Wu, H.; Lee, C. Self-powered triboelectric nanogenerator buoy ball for applications ranging from environment monitoring to water wave energy farm. *Nano Energy* **2017**, *40*, 203–213. [CrossRef]
18. Zhu, G.; Lin, Z.H.; Jing, Q.; Bai, P.; Pan, C.; Yang, Y.; Zhou, Y.; Wang, Z.L. Toward large-scale energy harvesting by a nanoparticle-enhanced triboelectric nanogenerator. *Nano Lett.* **2013**, *13*, 847–853. [CrossRef]
19. Zhao, X.; Wei, G.; Li, X.; Qin, Y.; Xu, D.; Tang, W.; Yin, H.; Wei, X.; Jia, L. Self-powered triboelectric nano vibration accelerometer based wireless sensor system for railway state health monitoring. *Nano Energy* **2017**, *34*, 549–555. [CrossRef]

20. Yu, J.; Hou, X.; He, J.; Cui, M.; Wang, C.; Geng, W.; Mu, J.; Han, B.; Chou, X. Ultra-flexible and high-sensitive triboelectric nanogenerator as electronic skin for self-powered human physiological signal monitoring. *Nano Energy* **2020**, *69*, 104437. [CrossRef]
21. Kim, S.; Towfeeq, I.; Dong, Y.; Gorman, S.; Rao, A.; Koley, G. P(VDF-TrFE) Film on PDMS Substrate for Energy Harvesting Applications. *Appl. Sci.* **2018**, *8*, 213. [CrossRef]
22. Lapčinskis, L.; Malnieks, K.; Linarts, A.; Blums, J.; Šmits, K.; Järvekülg, M.; Knite, M.; Šutka, A. Hybrid Tribo-Piezo-Electric Nanogenerator with Unprecedented Performance Based on Ferroelectric Composite Contacting Layers. *ACS Appl. Energ. Mater.* **2019**, *2*, 4027–4032. [CrossRef]
23. Han, M.; Zhang, X.-S.; Meng, B.; Liu, W.; Tang, W.; Sun, X.; Wang, W.; Zhang, H. r-Shaped Hybrid Nanogenerator with Enhanced Piezoelectricity. *ACS Nano* **2013**, *7*, 8554–8560. [CrossRef] [PubMed]
24. Ma, T.; Gao, Q.; Li, Y.; Wang, Z.; Lu, X.; Cheng, T. An Integrated Triboelectric–Electromagnetic–Piezoelectric Hybrid Energy Harvester Induced by a Multifunction Magnet for Rotational Motion. *Adv. Eng. Mater.* **2019**, *22*, 1900872. [CrossRef]
25. Gao, Y.J.; Leng, Y.G.; Fan, S.B.; Lai, Z.H. Performance of bistable piezoelectric cantilever vibration energy harvesters with an elastic support external magnet. *Smart Mater. Struct.* **2014**, *23*, 095003. [CrossRef]
26. Liu, H.; Lee, C.; Kobayashi, T.; Tay, C.J.; Quan, C. Investigation of a MEMS piezoelectric energy harvester system with a frequency-widened-bandwidth mechanism introduced by mechanical stoppers. *Smart Mater. Struct.* **2012**, *21*, 035005. [CrossRef]
27. Toyabur, R.M.; Salauddin, M.; Cho, H.; Park, J.Y. A multimodal hybrid energy harvester based on piezoelectric-electromagnetic mechanisms for low-frequency ambient vibrations. *Energy Convers. Manag.* **2018**, *168*, 454–466. [CrossRef]
28. Dhote, S.; Li, H.; Yang, Z. Multi-frequency responses of compliant orthoplanar spring designs for widening the bandwidth of piezoelectric energy harvesters. *Int. J. Mech. Sci.* **2019**, *157–158*, 684–691. [CrossRef]
29. Liu, H.; Koh, K.H.; Lee, C. Ultra-wide frequency broadening mechanism for micro-scale electromagnetic energy harvester. *Appl. Phys. Lett.* **2014**, *104*, 053901. [CrossRef]
30. Leland, E.S.; Wright, P.K. Resonance tuning of piezoelectric vibration energy scavenging generators using compressive axial preload. *Smart Mater. Struct.* **2006**, *15*, 1413–1420. [CrossRef]
31. Hajati, A.; Kim, S.-G. Ultra-wide bandwidth piezoelectric energy harvesting. *Appl. Phys. Lett.* **2011**, *99*, 083105. [CrossRef]
32. Stanton, S.C.; McGehee, C.C.; Mann, B.P. Reversible hysteresis for broadband magnetopiezoelastic energy harvesting. *Appl. Phys. Lett.* **2009**, *95*, 174103. [CrossRef]
33. Li, H.T.; Qin, W.Y. Prompt efficiency of energy harvesting by magnetic coupling of an improved bi-stable system. *Chin. Phys. B* **2016**, *25*, 110503. [CrossRef]
34. Cao, J.; Wang, W.; Zhou, S.; Inman, D.J.; Lin, J. Nonlinear time-varying potential bistable energy harvesting from human motion. *Appl. Phys. Lett.* **2015**, *107*, 143904. [CrossRef]
35. Gao, Y.; Leng, Y.; Javey, A.; Tan, D.; Liu, J.; Fan, S.; Lai, Z. Theoretical and applied research on bistable dual-piezoelectric-cantilever vibration energy harvesting toward realistic ambience. *Smart Mater. Struct.* **2016**, *25*, 115032. [CrossRef]
36. Liu, H.; Lee, C.; Kobayashi, T.; Tay, C.J.; Quan, C. Piezoelectric MEMS-based wideband energy harvesting systems using a frequency-up-conversion cantilever stopper. *Sens. Actuators A Phys.* **2012**, *186*, 242–248. [CrossRef]
37. Fu, H.; Zhou, S.; Yeatman, E.M. Exploring coupled electromechanical nonlinearities for broadband energy harvesting from low-frequency rotational sources. *Smart Mater. Struct.* **2019**, *28*, 075001. [CrossRef]
38. Stanton, S.C.; Mann, B.P.; Owens, B.A.M. Melnikov theoretic methods for characterizing the dynamics of the bistable piezoelectric inertial generator in complex spectral environments. *Phys. D* **2012**, *241*, 711–720. [CrossRef]
39. Bouhedma, S.; Zheng, Y.; Lange, F.; Hohlfeld, D. Magnetic Frequency Tuning of a Multimodal Vibration Energy Harvester. *Sensors* **2019**, *19*, 1149. [CrossRef]
40. Aldawood, G.; Nguyen, H.T.; Bardaweel, H. High power density spring-assisted nonlinear electromagnetic vibration energy harvester for low base-accelerations. *Appl. Energy* **2019**, *253*, 113546. [CrossRef]
41. Mann, B.P.; Sims, N.D. Energy harvesting from the nonlinear oscillations of magnetic levitation. *J. Sound Vib.* **2009**, *319*, 515–530. [CrossRef]
42. Chen, Y.; Pollock, T.E.; Salehian, A. Analysis of compliance effects on power generation of a nonlinear electromagnetic energy harvesting unit; theory and experiment. *Smart Mater. Struct.* **2013**, *22*, 094027. [CrossRef]
43. Lu, Z.; Wen, Q.; He, X.; Wen, Z. A Nonlinear Broadband Electromagnetic Vibration Energy Harvester Based on Double-Clamped Beam. *Energies* **2019**, *12*, 2710. [CrossRef]
44. Sun, S.; Dai, X.; Wang, K.; Xiang, X.; Ding, G.; Zhao, X. Nonlinear Electromagnetic Vibration Energy Harvester With Closed Magnetic Circuit. *IEEE Magn. Lett.* **2018**, *9*, 1–4. [CrossRef]
45. Xing, X.; Lou, J.; Yang, G.M.; Obi, O.; Driscoll, C.; Sun, N.X. Wideband vibration energy harvester with high permeability magnetic material. *Appl. Phys. Lett.* **2009**, *95*, 134103. [CrossRef]
46. Xing, X.; Yang, G.M.; Liu, M.; Lou, J.; Obi, O.; Sun, N.X. High power density vibration energy harvester with high permeability magnetic material. *J. Appl. Phys.* **2011**, *109*, 07E514. [CrossRef]
47. Ren, Z.; Zhao, H.; Liu, C.; Qian, L.; Zhang, S.; Zhao, J. Study the influence of magnetic force on nonlinear energy harvesting performance. *AIP Adv.* **2019**, *9*, 105107. [CrossRef]
48. Haroun, A.; Yamada, I.; Warisawa, S.i. Micro electromagnetic vibration energy harvester based on free/impact motion for low frequency–large amplitude operation. *Sens. Actuators A Phys.* **2015**, *224*, 87–98. [CrossRef]

49. Gu, Y.; Liu, W.; Zhao, C.; Wang, P. A goblet-like non-linear electromagnetic generator for planar multi-directional vibration energy harvesting. *Appl. Energy* **2020**, *266*, 114846. [CrossRef]
50. Fan, K.; Liu, S.; Liu, H.; Zhu, Y.; Wang, W.; Zhang, D. Scavenging energy from ultra-low frequency mechanical excitations through a bi-directional hybrid energy harvester. *Appl. Energy* **2018**, *216*, 8–20. [CrossRef]
51. Salauddin, M.; Rasel, M.S.; Kim, J.W.; Park, J.Y. Design and experiment of hybridized electromagnetic-triboelectric energy harvester using Halbach magnet array from handshaking vibration. *Energy Convers. Manag.* **2017**, *153*, 1–11. [CrossRef]
52. Askari, H.; Asadi, E.; Saadatnia, Z.; Khajepour, A.; Khamesee, M.B.; Zu, J. A hybridized electromagnetic-triboelectric self-powered sensor for traffic monitoring: Concept, modelling, and optimization. *Nano Energy* **2017**, *32*, 105–116. [CrossRef]

 *micromachines*

*Article*

# Modeling, Validation, and Performance of Two Tandem Cylinder Piezoelectric Energy Harvesters in Water Flow

Rujun Song [1,2], Chengwei Hou [1,3], Chongqiu Yang [1], Xianhai Yang [1], Qianjian Guo [1,*] and Xiaobiao Shan [3,*]

1   School of Mechanical Engineering, Shandong University of Technology, Zibo 255049, China; songrujun@sdut.edu.cn (R.S.); hcwking@163.com (C.H.); yangcq@sdut.edu.cn (C.Y.); yxh@sdut.edu.cn (X.Y.)
2   Shenzhen Research Institute of City University of Hong Kong, Shenzhen 518057, China
3   School of Mechatronics Engineering, Harbin Institute of Technology, Harbin 150001, China
*   Correspondence: guoqianjian@163.com (Q.G.); shanxiaobiao@hit.edu.cn (X.S.)

**Abstract:** This paper studies a novel enhanced energy-harvesting method to harvest water flow-induced vibration with a tandem arrangement of two piezoelectric energy harvesters (PEHs) in the direction of flowing water, through simulation modeling and experimental validation. A mathematical model is established by two individual-equivalent single-degree-of-freedom models, coupled with the hydrodynamic force obtained by computational fluid dynamics. Through the simulation analysis, the variation rules of vibration frequency, vibration amplitude, power generation and the distribution of flow field are obtained. And experimental tests are performed to verify the numerical calculation. The experimental and simulation results show that the upstream piezoelectric energy harvester (UPEH) is excited by the vortex-induced vibration, and the maximum value of performance is achieved when the UPEH and the vibration are resonant. As the vortex falls off from the UPEH, the downstream piezoelectric energy harvester (DPEH) generates a responsive beat frequency vibration. Energy-harvesting performance of the DPEH is better than that of the UPEH, especially at high speed flows. The maximum output power of the DPEH (371.7 μW) is 2.56 times of that of the UPEH (145.4 μW), at a specific spacing between the UPEN and the DPEH. Thereupon, the total output power of the two tandem piezoelectric energy harvester systems is significantly greater than that of the common single PEH, which provides a good foreground for further exploration of multiple piezoelectric energy harvesters system.

**Keywords:** piezoelectric energy harvester; tandem; energy harvesting; vortex-induced vibration; flowing water



**Citation:** Song, R.; Hou, C.; Yang, C.; Yang, X.; Guo, Q.; Shan, X. Modeling, Validation, and Performance of Two Tandem Cylinder Piezoelectric Energy Harvesters in Water Flow. *Micromachines* **2021**, *12*, 872. https://doi.org/10.3390/mi12080872

Academic Editors: Qiongfeng Shi and Huicong Liu

Received: 21 June 2021
Accepted: 23 July 2021
Published: 25 July 2021

**Publisher's Note:** MDPI stays neutral with regard to jurisdictional claims in published maps and institutional affiliations.



## 1. Introduction

Global environmental and energy policies stress the need to increase the share of renewable resources and to enhance the efficiency of energy conversion plants and commit to developing advanced solutions for power production [1–4]. In addition to the conversion of biogas to electricity, which has attracted many scholars, the use of piezoelectric materials to convert kinetic energy in the environment into electrical energy has also become a hot spot in electrical energy conversion. Fluid kinetic energy is one of the most important clean energy resources and is widely distributed. It is of great value to convert fluid kinetic energy to electricity for human use. There are several efficient methods for energy transformations, such as turbines for hydro-plants, windmills for wind farms, and so on. Both turbines and windmills are suitable for large-scale power generation, which is inconvenient for portable, wearable, low-energy device. The piezoelectric effect can be applied for energy harvesting [5–7] and ultrasonic transduction [8–11]. Especially, piezoelectric energy harvesters (PEHs) have been widely used for energy harvesting from ambient vibration [12–16]. Hence, the piezoelectric conversion mechanism can be used to capture energy from fluid kinetic energy, which has attracted much attention among scholars [17–22].

Flow-induced vibration (FIV) and biogas are renewable and alternative sources of energy [23–25]. FIV is a common natural phenomenon, such as vortex-induced vibration of smokestacks and power transmission lines, fluttering flags in airflow, and the flow-induced vibration of seaweed and submarine cables in ocean currents. Biogas and biomass gas (syngas) have been used to feed solid oxide fuel cells [26,27]. Bochentyn et al. [28] proposed using praseodymium and samarium co-doped with ceria as an anode catalyst for DIR-SOFC fueled by biogas, and investigated the catalytic performance. De Lorenzo et al. [29,30] investigated the electrical and thermal analysis of an intermediate-temperature solid oxide fuel cell system fed by biogas. Energy harvesting from the above FIV phenomenon will be an effective method for converting fluid kinetic energy to electric energy, which can be regarded as wake-induced vibration (WIV) [31–34], vortex-induced vibration [35–40], fluttering [41–44] and galloping [45–48], etc. Particularly, PEHs based on the WIV have been extensively investigated in the past decade years. Taylor et al. [49] and Allen et al. [50] first proposed an eel-shaped PEH, based on the WIV to convert water kinetic energy into electric energy. Akaydin et al. [51] studied an airflow PEH by WIV and found that the maximum power output could be obtained when the tip of the piezoelectric beam was located downstream about twice the cylinders' diameter. Weinstein et al. [52] proposed that PEH caused by the awakening of the upstream cylinder, achieving output power of 3 mW at air velocity of 5 m/s. Li et al. [53] conducted a numerical study on the feasibility of the blunt body obtaining energy from steady state flow when $Re = 100$ and found that the performance was closely related to the local vortex dissipation and pressure gradient of large vortex in the wake near the PEH. Yu et al. [54] proposed four structures of cylindrical and PEH membranes. The beat dynamics of PEH membranes were extensively studied, which is helpful to comprehensively understand the energy collection process of elasticity, kinetic energy, and continuous energy transfer. It can be inferred that an enhanced performance of PEH can be achieved by using the WIV phenomenon.

Inspired by the WIV, some researchers have proposed a combined piezoelectric energy harvesting system to improve space utilization and output power performance [55]. Li et al. [56] investigated the wake effect of the two tandem cylinders and found that the aerodynamic forces of the two cylinders were largely determined by the spacing ratio between them. Abdelkefi et al. [57] studied two tandem PEHs in airflow. It was found that when the distance between two PEHs exceeds a critical value, wake galloping of the upstream PEH would occur, and this wake galloping can improve the average output power of the total energy harvesting system. Hu et al. [58] evaluated the performance of twin, adjacent PEHs, based on mutual interference, under wind excitation, and found that the total output power of the two wind energy harvesters could achieve up to 2.2 times that of two single harvesters at the optimal relative position. All above works indicate that the PEHs with tandem arrangement can improve the energy-harvesting capacity of the overall system. Therefore, Shan et al. [59] proposed an energy-harvesting system with two identical piezoelectric energy harvesters in a tandem configuration and investigated the energy-harvesting performance by experimentation. However, the vibration response mechanism of fluid–structure interaction and the relationship between the output power of the harvester and flow field distribution were not studied. Hence, this work proposes a simulation method to study the performance of two tandem cylinder piezoelectric energy harvesters thoroughly and systematically. The validity of the proposed simulation method is verified through the experimental results. Through the simulation, the vibration response and energy-harvesting performance under wake excitation, including the variation rules of vibration frequency, vibration amplitude, power generation, and the distribution of flow field, can be easily obtained.

## 2. Physical Model and Simulation Method

Figure 1 shows the physical model of a two tandem PEHs system. Each PEH consisted of a piezoelectric beam and a cylinder. The piezoelectric beam was fixed on the upper end and consisted of a piezoelectric layer (PZT-5H) and a substrate layer. The cylinder was attached to the free end and vertically immersed into incoming water. $U$ is the water velocity. The water flowed over the cylinders and stimulated the PEHs to vibrate periodically. The energy harvesting system can be equivalent into two tandem independent single degree of freedom PEHs with separation spacing $W$. As shown in Figure 1, $u$ is the vibration displacement of the PEH. $M_{eq}$, $K_b$ are the equivalent mass and equivalent stiffness, respectively. $C$ is the damping coefficient of the single PEH. $C_p$ is the equivalent capacitance of the PEH. $\Theta$ is the electromechanical coupling coefficient of mechanical vibration and electrical output. $V$ is the generated voltage across the electrical load resistance $R$.

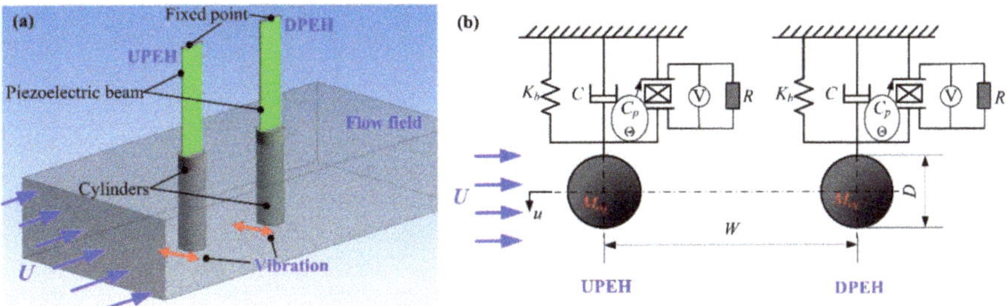


**Figure 1.** Schematic diagram of two tandem PEHs system: (**a**) three-dimensional system; (**b**) equivalent model of two tandem PEHs.

According to the equivalent model in Figure 1b, each of the two PEHs can be described as an individual-equivalent single-degree-of-freedom model, written as

$$\begin{cases} M_{eq}\ddot{u} + C\dot{u} + K_b u + \Theta V = F_f \\ \frac{V}{R} + C_p \dot{V} - \Theta \dot{u} = 0 \end{cases} \tag{1}$$

where, $F_f$ is the flow-induced force, which should be obtained by the computational fluid dynamics (CFD) method (Fluent® software) due to the complex interactions between the two tandem cylinders in water. $C$ is the damping equaling to $2\omega_n\zeta$, and the $\zeta$ is the damping ratio and can be obtained by free decay oscillation method in experimental text. The second equation is based on the Kirchhoff's law, which express that the sum of all the currents entering a node is equal to the sum of all the currents leaving the node. A key point of the measurement in the experiment is that the cylinder must be immersed into water when the free decay oscillation occurs. As a result, the added fluid damping of cylinder must be taken into account.

Figure 2a illuminates the schematic of the individual PEH. The length of piezoelectric beam and cylinder are $L_b$, $L_c$. The thickness of piezoelectric and substrate layer are $h_p$, $h_s$, respectively. The width of piezoelectric beam is $b$. The depth of immersion into water flow of the PEH is $L_f$. Point $O$ is the center of flow-induced force action on cylinder. Figure 2b shows the finite element analytical model of the PEH. Elements SOLID5, SOLID45, and CIRCU94, in ANSYS® 10, were used for finite element simulation of the piezoelectric layer, substrate lay, and load resistance circuit, respectively. The sign of the piezoelectric constant was applied for indicating the polarization direction of the PZT-5H. The node on the upper piezoelectric layer surface was named common node "1", and the node on the lower surface was named common node "2". Then the load resistance $R$ can be connected between the common nodes "1" and "2".

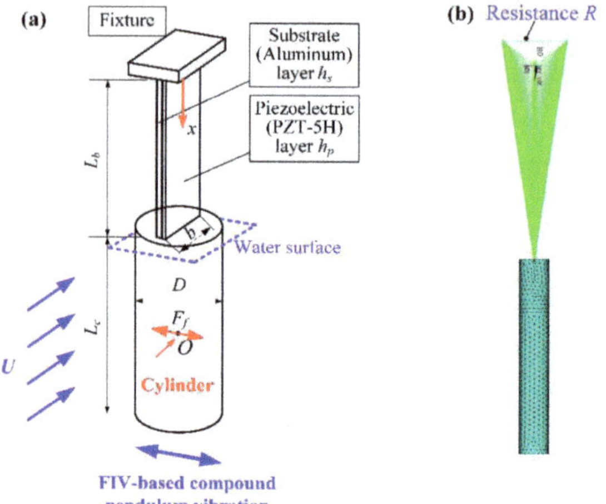



**Figure 2.** Physical model of the individual PEH: (**a**) Dimension diagram, (**b**) Finite element analytical model.

According to dimension of PEH in Figure 2a, the deflection $w_b(x, t)$ of the cantilever beam can be expressed as

$$w_b(x,t) = \frac{F_f x^2}{6EI}(3L_b - x) + \frac{M_{Force}x^2}{2EI} \qquad x \leq L_b \qquad (2)$$

where $EI$ is the stiffness of piezoelectric cantilever beam, expressed as

$$EI = b\left[\frac{E_p\left(h_c^3 - h_b^3\right) + E_s\left(h_b^3 - h_a^3\right)}{3}\right] \qquad (3)$$

where, $h_a$, $h_b$, and $h_c$ are expressed as

$$\begin{aligned}
h_a &= -\frac{h_s}{2} - \frac{(h_p + h_s)E_p h_p}{2(E_p h_p + E_s h_s)} \\
h_b &= \frac{h_s}{2} - \frac{(h_p + h_s)E_p h_p}{2(E_p h_p + E_s h_s)} \\
h_c &= \frac{h_s}{2} + h_p - \frac{(h_p + h_s)E_p h_p}{2(E_p h_p + E_s h_s)}
\end{aligned} \qquad (4)$$

where, $E_p$, $E_s$ are the Young modulus of the piezoelectric and substrate materials.

The moment $M_{Force}$ acts on the tail of cantilever beam is expressed as

$$M_{Force} = F_f\left(L_c - \frac{L_f}{2}\right) \qquad (5)$$

The displacement of terminal cylinder $w(x, t)$ is

$$w(x,t) = w_b(L_b, t) + w_b'(L_b, t)(x - L_b) \qquad L_b \leq x \leq L_b + L_c \qquad (6)$$

where, $w_b'$ is the first derivative of $x$. Then equivalent stiffness $K_b$ can be obtained by the following equation, using Hooke's law,

$$K_b = \frac{F_f}{w(L_{bo}, t)} \qquad (7)$$

where, $L_{bo} = L_b + L_c - L_f/2$, and the $K_b$ is obtained at last.

The total kinetic energy of the PEH (TPEH) is the sum of kinetic energy of the piezoelectric layer and substrate layer ($T_{beam}$), kinetic energy of cylinder ($T_{cylinder}$), and kinetic energy of its added fluid ($T_{fluid}$), expressed as follow

$$T_{PEH} = T_{beam} + T_{cylinder} + T_{fluid} \tag{8}$$

where,

$$
\begin{aligned}
T_{beam} &= \tfrac{1}{2}\int_{V_p}\rho_p\left[\frac{\partial w_b(x,t)}{\partial t}\right]^2 dV_p + \tfrac{1}{2}\int_{V_s}\rho_s\left[\frac{\partial w_b(x,t)}{\partial t}\right]^2 dV_s \\
T_{cylinder} &= \tfrac{1}{2}\int_{L_b}^{L_b+L_c}\frac{M_c}{L_c}\left[\frac{\partial w(x,t)}{\partial t}\right]^2 dx \\
T_{fluid} &= \tfrac{1}{2}\int_{L_b+L_c-L_f}^{L_b+L_c}\frac{M_{cf}}{L_f}\left[\frac{\partial w(x,t)}{\partial t}\right]^2 dx
\end{aligned}
\tag{9}
$$

where, $\rho_s$ is the density of substrate material, $\rho_p$ is the density of piezoelectric material, $M_c$ is the mass of cylinder, and $M_{cf}$ is the fluid added mass, written as

$$M_{cf} = C_M\frac{D^2}{4}\pi L_f\rho_f \tag{10}$$

where $C_M$ is the additional mass coefficient, $C_M = 1$.

Meanwhile, the kinetic energy TPEH can be re-expressed as

$$T_{PEH} = \frac{1}{2}M_{eq}\left[\dot{w}(L_{bo},t)\right]^2 \tag{11}$$

and the $M_{eq}$ can be obtained based on Equations (8) and (11). Thus the equivalent mass $M_{eq}$ can be obtained by the equation:

$$M_{eq} = \left[\begin{array}{l}\int_{V_p}\rho_p\left[\frac{\partial w_b(x,t)}{\partial t}\right]^2 dV_p + \int_{V_s}\rho_s\left[\frac{\partial w_b(x,t)}{\partial t}\right]^2 dV_s \\ + \int_{L_b}^{L_b+L_c}\frac{M_c}{L_c}\left[\frac{\partial w(x,t)}{\partial t}\right]^2 dx + \int_{L_b+L_c-L_f}^{L_b+L_c}\frac{M_{cf}}{L_f}\left[\frac{\partial w(x,t)}{\partial t}\right]^2 dx\end{array}\right] / \left[\dot{w}(L_{bo},t)\right]^2$$

The $\Theta$ of electromechanical coupling coefficient of the single PEH in energy harvesting system can be expressed as

$$\Theta = \sqrt{(w_{noc}^2 - w_{nsc}^2)M_{eq}C_p} \tag{12}$$

where, $\omega_{noc}$ is open circuit resonance frequency, and $\omega_{nsc}$ is short circuit resonance frequency, which will be obtained by $R_0 = 0$ and $R_0 \rightarrow \infty$ in ANSYS® software, respectively. Equivalent capacitance $C_p$ can be

$$C_p = \frac{\varepsilon_{33}^S bL_b}{h_p} \tag{13}$$

where $\varepsilon_{33}^S$ is the dielectric constant of piezoelectric material.

Furthermore, the output power of the PEH can be expressed as

$$P = \frac{1}{T}\int_0^T \frac{V^2(t)}{R}dt \tag{14}$$

where $T$ is the alternating period of output voltage.

The simulation results can be obtained by Equations (1) and (14) using the User Defined Function (UDF) in Fluent® software. The simulation process can be summarized as:

(1) Flow-induced force $F_f$ can be obtained by Fluent® software at the cylinder position in flow field at time $t$.

(2) The vibration displacement $u$, vibration velocity $\dot{u}$, vibration acceleration $\ddot{u}$ of the PEH, and electrical parameters $V(t)$, $P$ can be obtained by both Equations (1) and (14).

(3) The cylinder position and computational grid will be updated based on the PEH displacement $u$ in step 2.

(4) One step $\Delta t$ is finished and next step $t + \Delta t$ will proceed. The performances of two tandem PEHs can then be obtained at last.

Figure 3 illuminates the schematic of the flow field, including the whole grid arrangement and partial grid diagram of two tandem PEHs. The dynamic grid was applied in this CFD process. The domain of flow field grid was $20D \times (40D + W)$, where $D$ is the cylinder diameter and $W$ is the spacing of the two tandem PEHs. The spacing of inlet to upstream piezoelectric energy harvester (UPEH) was 10 times that of $D$. The spacing of outlet to downstream piezoelectric energy harvester (DPEH) was 30 times that of $D$. In order to ensure the grid quality near the cylinders during the simulation process, the computational domain was divided into three parts, as shown in Figure 3a. The first part was the static grid. The second part was the dynamic grid, where the third part could move with the two tandem PEHs during the CFD process to reduce grid distortion, as shown in Figure 3b.

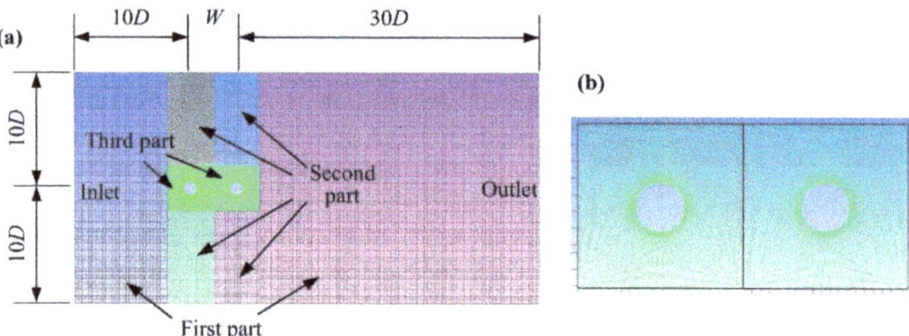


**Figure 3.** Schematic of the flow field grid of the tandem arrangement of energy harvesting system: (**a**) grid diagram of flow field; (**b**) partial grid diagram of third part.

## 3. Experimental Setup and Simulation Verification

Figure 4 shows the experimental platform and the prototype setup. The experiment platform was made up of an open channel, fixtures, and measurement system. In order to obtain a steady flow, some sections in the open channel, such as the contraction section, damping screens, and a cellular device were set up to reduce stream turbulence. The speed of the water was controlled by a pump driven by a frequency converter. The fixtures were used to fix the two tandem PEHs, the spacing $W$ was changed by relative position of the fixtures. The measurement system was composed of an adjustable resistor, a Data Acquisition system (DAQ) from National Instruments (NI), and a computer. The output voltage of the resistor was monitored by DAQ and displayed on the computer.

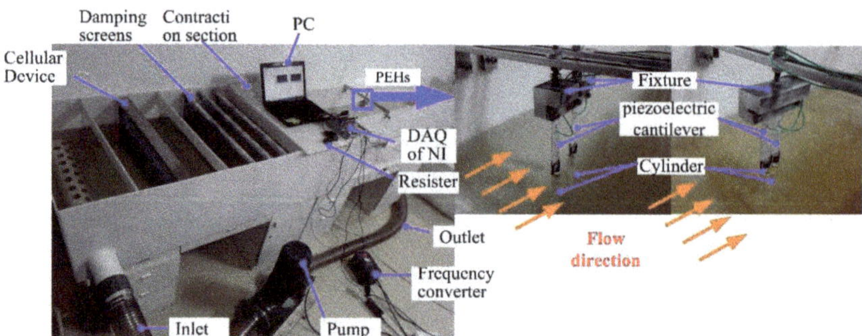


**Figure 4.** Energy harvesting experimental platform and prototypes of two tandem PEHs.

Figure 5 shows the output voltage evolution curve of the upstream and downstream PEHs with time when the spacing ratio $W/D$ is 3.3 at the flow velocity of 0.23 m/s, 0.31 m/s and 0.37 m/s, respectively. The dimension and parameters of the two tandem PEHs are listed in Table 1. The comparison shows that when the flow velocity is 0.23 m/s, the output voltage of the upstream PEH was greater than that of the downstream PEH. At a flow velocity of 0.37 m/s, the output voltage of the upstream PEH was smaller than that of the downstream PEH. By comparing Figure 5a,c,e, it can be found that the output voltage of the upstream PEH at the velocity of 0.31 m/s is higher than that at the flow velocity of 0.23 m/s and 0.37 m/s, respectively. By comparing Figure 5b,d,f, it can be seen that the output voltage of the downstream PEH increases with the increase in velocity. It can be seen that under the tandem arrangement of the two PEHs, due to the strong coupling effect, the performance of the two PEHs will change greatly, especially the downstream one.

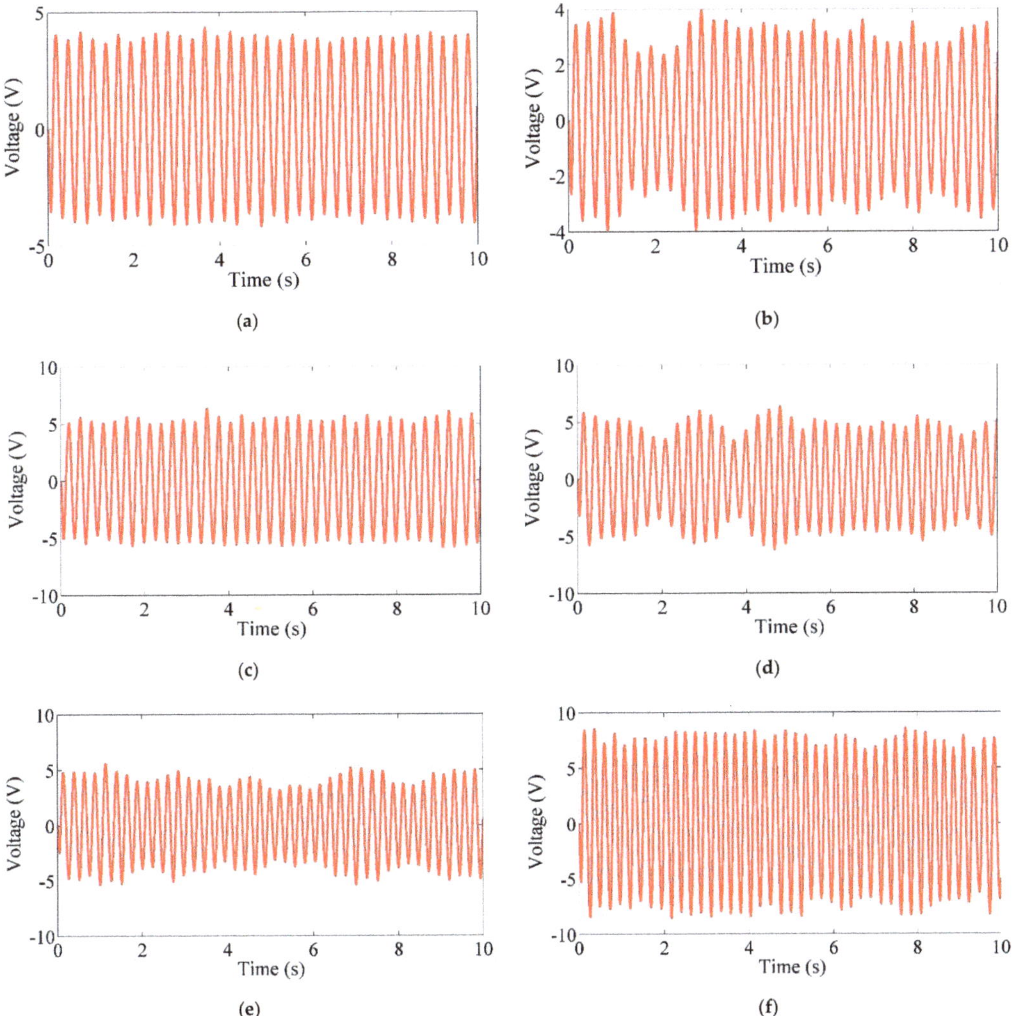

**Figure 5.** Output voltage of two tandem PEHs versus time for $W/D = 3.3$. (**a**) Output voltage of UPEH at $U = 0.23$ m/s; (**b**) Output voltage of DPEH at $U = 0.23$ m/s; (**c**) Output voltage of UPEH at $U = 0.31$ m/s; (**d**) Output voltage of DPEH at $U = 0.31$ m/s; (**e**) Output voltage of UPEH at $U = 0.37$ m/s; (**f**) Output voltage of DPEH at $U = 0.37$ m/s.

**Table 1.** Dimension and material properties of the PEHs.

| Parameters | Values |
| --- | --- |
| density of PZT-5H $\rho_p$ (kg/m$^3$) | 7386 |
| density of aluminum layer $\rho_p$ (kg/m$^3$) | 2700 |
| fluid density $\rho_p$ (kg/m$^3$) | 1000 |
| Young modulus of PZT-5H $E_p$ (GPa) | 59.77 |
| Young modulus of aluminum layer $E_s$ (GPa) | 71.7 |
| length $\times$ width $\times$ thickness of piezoelectric layer ($L_b \times b \times h_p$) (mm) | $80 \times 20 \times 0.2$ |
| length $\times$ width $\times$ thickness of substrate layer ($L_b \times b \times h_s$) (mm) | $80 \times 20 \times 0.2$ |
| piezoelectric constant $e_{31}$ (C/m$^2$) | $-13.74$ |
| dielectric constant $\varepsilon_{33}^s$ (F/m) | $4.178 \times 10^{-8}$ |
| load resistance $R$ (k$\Omega$) | 100 |
| Cylinder diameter $D$ (mm) | 12 |
| length of cylinder immersed in water $L_f$ (mm) | 55 |
| first mode damping ratio $\zeta$ | 0.0499 |
| open circuit resonance frequency $\omega_{noc}$ (rad) | 26.25 |
| short circuit resonance frequency $\omega_{nsc}$ (rad) | 25.5 |

In order to research the energy-harvesting performance and verify the effectiveness of simulation method, Figure 6 illustrates the output power comparison between simulation and experimental results of the two tandem PEHs when the spacing ratio $W/D$ is 3.3, 4.58, and 6.25, respectively. As can be seen from Figure 6a–c, when the spacing was smaller, the coupling effect was stronger, and the output power of the UPEH was slightly improved. With the increase in spacing ratio, the coupling effect was weakened and the energy harvesting performance of the UPEH decreased and gradually tended to resemble the performance of a single PEH. The experimental results showed that peak output power (145.4 µW of UPEH and 371.7 µW of DPEH) was generated at the velocities of 0.31 m/s and 0.41 m/s when $W/D$ = 3.3, respectively. The maximum output power of DPEH was 2.56 times of that of the UPEH. When $W/D$ = 6.25, the maximum output power of the UPEH was 117.9 µW and the maximum output power of the DPEH was 239.9 µW. It can be seen that the output performance of DPEH was better than that of UPEH, mainly occurred at the high flow velocity (>0.31 m/s). Especially, the output power of DPEH (371.7 µW) was 12.8 times that of UPEH (29.2 µW) when velocity was 0.41 m/s and $W/D$ = 3.3. It can be summarized that the coupling effect of each PEH decreased with the increase in the spacing ratio of $W/D$. Meanwhile, the wake effect of the DPEH from the UPEH was gradually decreased, and the disturbance effect on UPEH from the DPEH was also weakened. It's worth noting that the comparison between the theoretical results and the experimental data is better for high $W/D$ values. This is because, in the experiment, the smaller the distance, the more serious the impact of the coupled vibration. The mutual disturbance is bound to cause disturbances in the flow field, aggravate wall disturbances, and change the depth of the cylinder's water. In the theoretical analysis, the wall of the flow field is far enough away and the immersion depth of the two cylinders is assumed to be constant. This leads to smaller distances producing greater error between the simulation analysis results and the experimental results. The experimental results also show that the output power of UPEH and DPEH increased to the maximum first, and then decreased with the increase of the flow velocity, and the optimal velocity of the maximum power output could be obtained. In addition, the energy harvesting performance of a single piezoelectric energy harvester (SPEH) is given in Figure 6d. It can be noted that the simulation flow velocity corresponding to the maximum output power is lower than the experimental results. This is mainly because when the vortex-induced resonance occurs, the column immersion depth is slightly reduced, and the additional mass of water to the cylinder is slightly reduced too, but the immersion depth is assumed to be unchanged during the simulation, so that the natural frequency of the energy vibrator during simulation is slightly lower, which causes the flow velocity corresponding to the vortex-induced resonance to be slightly smaller than the experimental result. By comparing Figure 5a–c, it can be found that the power generation capacity of the UPEH and DPEH were greater than that of the SPEH. Therefore, it can be seen that the two tandem PEHs can improve the energy-harvesting performance of the system. What's more, the overall results of the experiment and simulation are mild, and the error is within the allowable range. The main sources of error can be summarized as: in the experiment, the left and right swing caused a certain change in the depth of the cylinder's water inlet. In the simulation, it is treated as quantitative, and some assumptions in the simplification process of the two-dimensional model will also have certain contrast errors.

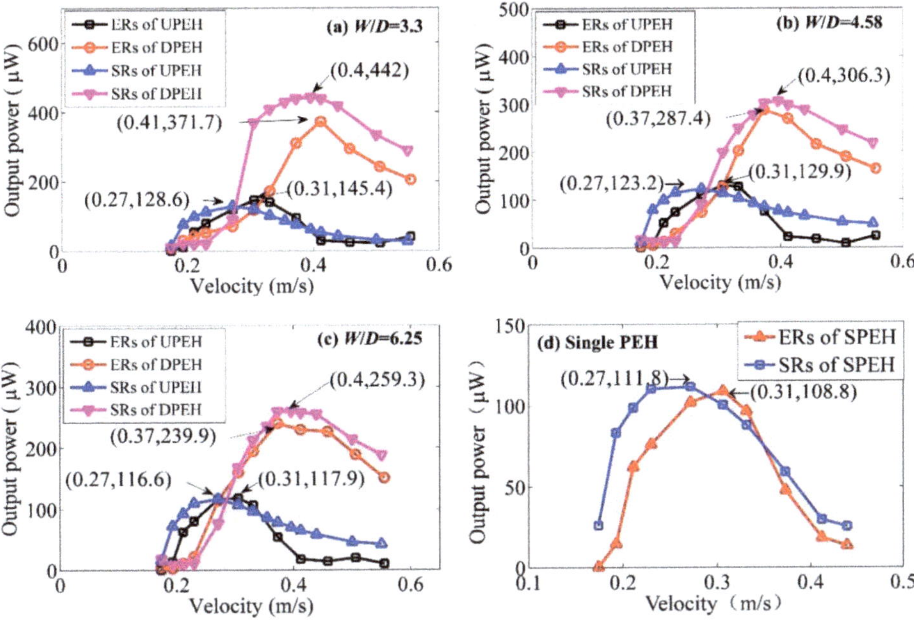

**Figure 6.** Output power of simulation results (SRs) and experimental results (ERs) of two tandem PEHs (**a**–**c**) and single PEH (**d**).

Therefore, the simulation results were consistent with the experimental results, as shown in Figure 6; the simulation method was verified.

## 4. Simulation Results and Discussion

In order to further investigate the performance of the two tandem PEHs, Figure 7a,b shows changes of vibration frequencies versus velocity. It can be concluded that frequency increases with velocity as a whole. However, as for the DPEH, the vibration performance is complicated. As shown in Figure 7a, the frequency increases with the increase of water velocity when water velocity ($U$) is less than 0.211 m/s, and then decreases with the increase of water velocity when 0.211 m/s < $U$ < 0.23 m/s, and lastly increases with the increase of water velocity when $U$ > 0.23 m/s. When the $W/D$ = 12.5, the vibration frequency of DPEH was smaller than that of UPEH at low-speed flow, as shown in Figure 7b. Figure 7c–f are plotted for when $U$ = 0.272 m/s. Due to the coupling effect of the vortex-induced vibration (VIV) itself and the wake-induced vibration (WIV) from the UPEH, there was inevitable unbalanced vibration in the DPEH, with small vibration frequency differences, such as the beat vibration signal, containing two frequency components of dominant frequency and non-dominant frequency when $W/D$ = 6.25, which can be found that Figure 7c,d. When $W/D$ = 12.5, the coupling effect becomes light, so that the frequency is unique as shown in Figure 7e,f.

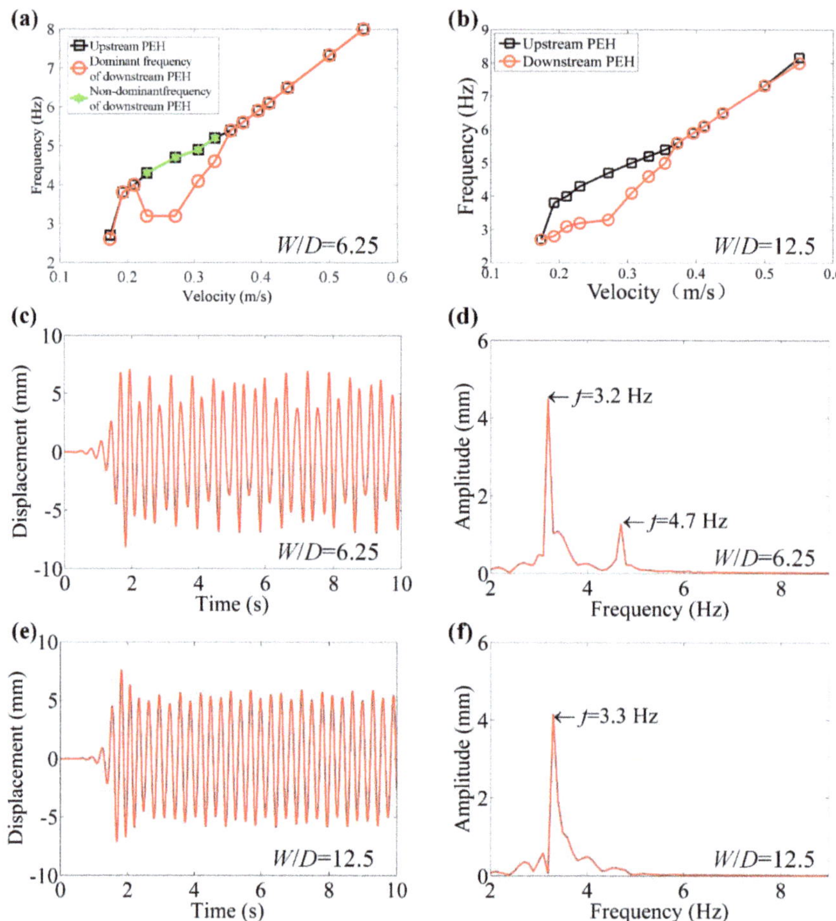

**Figure 7.** Vibration frequencies of the PEHs versus velocity when $W/D = 6.25$ and $W/D = 12.5$. (**a**) Vibration frequencies when $W/D = 6.25$. (**b**) Vibration frequencies when $W/D = 12.25$. (**c**) Vibration response when $W/D = 6.25$ and $U = 0.272$ m/s. (**d**) Fast Fourier transform (FFT) when $W/D = 6.25$ and $U = 0.272$ m/s. (**e**) Vibration response when $W/D = 12.25$ and $U = 0.272$ m/s. (**f**) FFT when $W/D = 12.25$ and $U = 0.272$ m/s.

In order to further analyze the vibration performance of the two tandem PEHs system, the vibration amplitudes of the upstream and downstream PEH with spacing ratio $W/D$ and water speed were numerically discussed, specifically. As seen in Figure 8, Part I is the area of water velocity and Part II is the area of high-speed water velocity. As for the UPEH, the vibration amplitude was not mainly affected by the spacing ratio $W/D$, but rather by the water velocity. The vibration amplitude of the UPEH first increased and then decreased with the increase in velocity, and reached its maximum value when the velocity was about 0.27 m/s. As for the DPEH, the vibration amplitude first decreased to the minimum value when the velocity was small, as shown in Part I in Figure 8b. This phenomenon was attributed to the flow field transition of the shear layer around the DPEH; the vibration response was weakened during the transformation of the water flow from laminar to turbulent. With further increase in the water velocity, the amplitude of vibration on the DPEH increased to its maximum point, and then gradually decreased to a stable value, as shown in Figure 8b. Due to the excitation effect of upstream vortex shedding from the UPEH, the vibration response of DPEH was obviously greater than that of UPEH, especially at the high water speed, as shown in Part II in Figure 8a,b respectively.

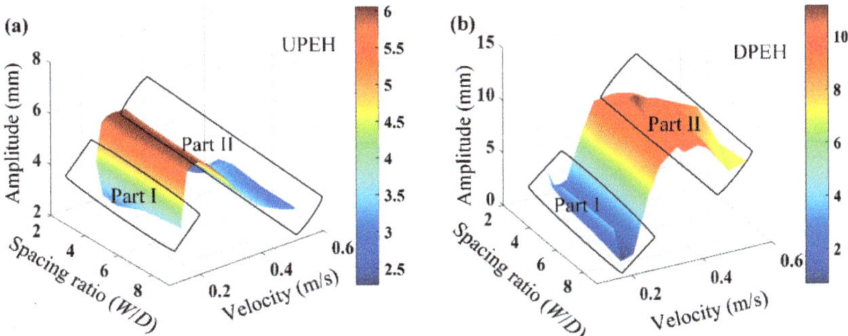

**Figure 8.** Comparison of vibration amplitudes of the upstream and downstream PEH: (**a**) UPEH, (**b**) DPEH.

In order to study the energy-harvesting performance of the system, the relationship among the output power of tandem PEHs, spacing ratio $W/D$, and water velocity are illustrated in Figure 9. It can be found that the output power of UPEH increased first and then decreased with the increase in water speed. The maximum output power was obtained at the speed of 0.27 m/s, at which point vortex-induced resonance occurred, combining with the vibration responses in Figure 8a. As for the DPEH in Figure 9b, the relationship between $W/D$ and the energy-harvesting performance of the DPEH was the same as that of the UPEH. Comparing with the UPEH, the output power of the DPEH decreased slowly with the increase in velocity, due to wake stimulation from upstream, which was different from the performance of the UPEH. In addition, it can be concluded that in the range of low water velocity, the output power of the UPEH was higher than that of the DPEH, but in the range of high water velocity, the energy-harvesting performance of the DPEH was obviously better than that of the UPE,H due to wake stimulation by vortex shedding, which is in accordance with the vibration response results in Figure 8.

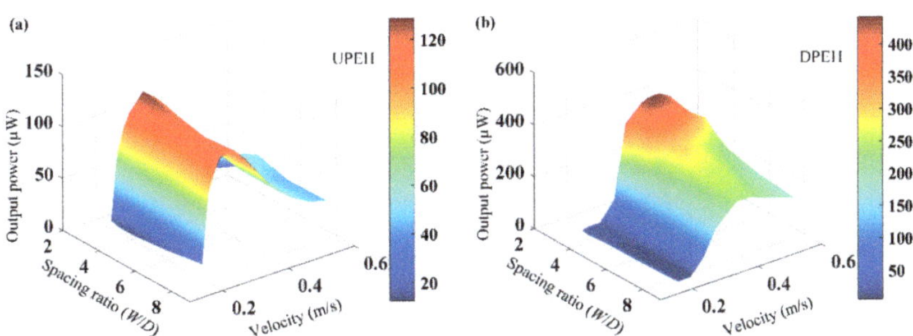

**Figure 9.** Energy harvesting performance of the two tandem PEHs versus spacing ratio $W/D$ and water velocity: (**a**) UPEH, (**b**) DPEH.

Further analyzing the coupling effect of the two tandem piezoelectric energy harvester, Figure 10 shows the velocity flow field contour in $x$ direction with different spacing ratios and water velocities. It can be seen from Figure 10a,c that at a low flow speed ($U = 0.19$ m/s), a long low-speed flow region could be formed in the downstream after the water flowed through the upstream PEH (cylinder). At this time, the downstream PEH would be in the low-speed flow region. Therefore, the vibration response and power generation performance of the downstream PEH were weak at the low-speed water flow, as shown in Figure 9b. At the same time, the vibration response of the upstream PEH was similar to that of a single PEH at low speed flow, because the downstream PEH had a weak vibration response and its influence on the upstream PEH was negligible. According to Figure 10b,d, when the flow velocity was high ($U$ = 0.41 m/s), the downstream PEH was located on the vortex path where the upstream PEH fallen off, and the surrounding vortices were relatively strong. Under stimulation of the vortex, the vibration response and power output performance of the downstream PEH would be greatly improved. In addition, as the vortex moved forward with the flow field, the vortex diffused and its turbulence strength decreased, so the vibration response and energy harvesting performance of the downstream PEH gradually decreased with the increase in spacing ratio.

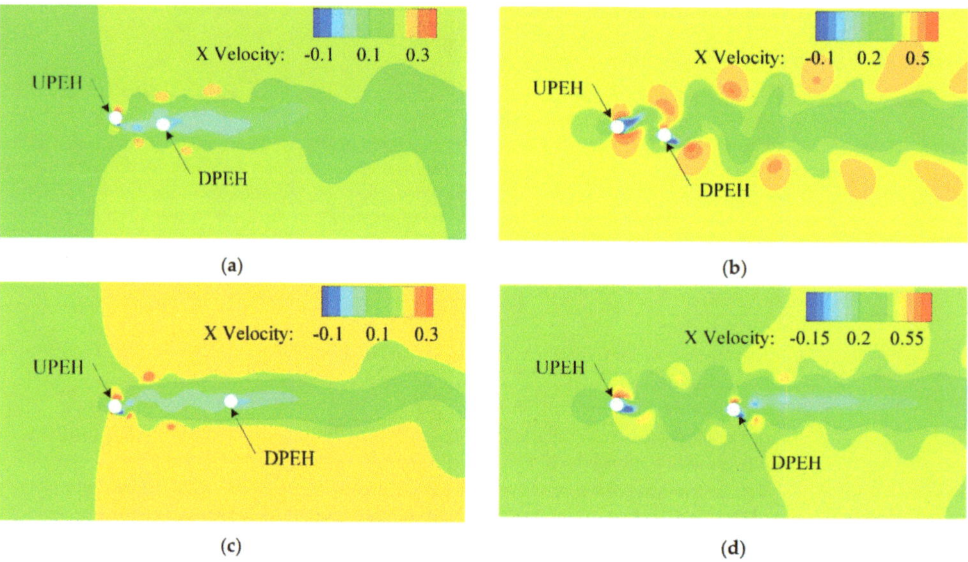


**Figure 10.** Flow field distribution of two tandem PEHs for various $W/D$ and water velocities. (**a**) Flow field distribution at $W/D$ = 3.33, $U$ = 0.19 m/s; (**b**) Flow field distribution at $W/D$ = 3.33, $U$ = 0.41 m/s; (**c**) Flow field distribution at $W/D$ = 3.33, $U$ = 0.19 m/s; (**d**) Flow field distribution at $W/D$ = 8.33, $U$ = 0.41 m/s.

## 5. Conclusions

In this paper, a tandem arrangement of piezoelectric energy harvesters was studied to scavenge the water flow vibration energy through simulation modeling and experimental validation. Through the simulation analysis, the variation rules of vibration frequency, vibration amplitude, power generation, and the distribution of flow field were obtained. The experiment results verify the accuracy of the simulation results. The effects of spacing ratio and water flow velocity on the vibration response and the output power of two tandem PEHs were studied numerically and experimentally. It could be concluded that the vibration response and energy-harvesting performance of the two tandem PEHs were enhanced due to the coupling effect induced by upstream vortex vibration. The vibration response of UPEH was vortex-induced vibration, and maximum power could be achieved at the resonant point of the PEHs and the vortex flow. When the spacing was small, the performance was enhanced due to enhancement of the DPEH. The vibration response of the DPEH was raised under the stimulation of the vortices shedding from the UPEH, especially under high-speed water flow; the energy-harvesting performance of the DPEH as better than that of the UPEH accordingly. However, when the flow speed was low, it was noticeable that the superiority of the DPEH energy harvesting was not remarkable and the power output of the UPEH was better than that of the DPEH. The results show that the energy-harvesting performance of the tandem PEHs in flowing water could be significantly improved, especially at the high speed flow region, which provides good support for further exploration of energy harvesting systems with multi-piezoelectric harvesters, parallel or serially arranged.

**Author Contributions:** R.S. contributed to the whole work including design, fabrication, testing of the device, and writing—original draft preparation; C.H. contributed to the simulation analysis and writing—review and editing; C.Y. contributed to fabrication and simulation analysis; X.Y. contributed to the testing guidance; Q.G. contributed to the writing—review and editing and X.S. contributed to the research idea and gave guidance to the work. All authors have read and agreed to the published version of the manuscript.

**Funding:** This research was funded by the National Natural Science Foundation of China (Grant numbers 51705296, 51875116 and 52075306), Key Research and Development Project of Shandong Province under Grant 2019GGX104033, Shenzhen Science and Technology Innovation Committee under Grant No. JCYJ20200109143206663, Key Research and Development Project of Zibo City under Grant No. 2020SNPT0088 and Young Innovative Talents Introduction and Training Program Project of Shandong Provincial Department of Education.

**Acknowledgments:** The authors are also grateful to the experimental support of Harbin Institute of Technology.

**Conflicts of Interest:** The authors declare no conflict of interest.

# References


1. Mehr, A.S.; Mahmoudi, S.M.S.; Yari, M.; Chitsaz, A. Thermodynamic and exergoeconomic analysis of biogas fed solid oxide fuel cell power plants emphasizing on anode and cathode recycling: A comparative study. *Energy Convers. Manag.* **2015**, *105*, 596–606. [CrossRef]
2. Manenti, F.; Pelosato, R.; Vallevi, P.; Leon-Garzon, A.R.; Dotelli, G.; Vita, A.; Lo Faro, M.; Maggio, G.; Pino, L.; Aricò, A.S. Biogas-fed solid oxide fuel cell (SOFC) coupled to tri-reforming process: Modelling and simulation. *Int. J. Hydrogen Energy* **2015**, *40*, 14640–14650. [CrossRef]
3. Erturk, A.; Inman, D.J. Issues in mathematical modeling of piezoelectric energy harvesters. *Smart Mater. Struct.* **2008**, *17*, 065016. [CrossRef]
4. Shan, X.; Tian, H.; Chen, D.; Xie, T. A curved panel energy harvester for aeroelastic vibration. *Appl. Energy* **2019**, *249*, 58–66. [CrossRef]
5. Huang, D.; Zhou, S.; Yang, Z. Resonance Mechanism of Nonlinear Vibrational Multistable Energy Harvesters under Narrow-Band Stochastic Parametric Excitations. *Complexity* **2019**, *2019*, 1–20. [CrossRef]
6. Huang, D.; Zhou, S.; Han, Q.; Litak, G. Response analysis of the nonlinear vibration energy harvester with an uncertain parameter. *J. Multi-Body Dyn.* **2019**, *82*, 376–381. [CrossRef]
7. Zhou, S.; Zuo, L. Nonlinear dynamic analysis of asymmetric tristable energy harvesters for enhanced energy harvesting. *Commun. Nonlinear Sci. Numer. Simul.* **2018**, *61*, 271–284. [CrossRef]
8. Chen, W.; Liu, Y.; Yang, X.; Liu, J. Ring-type traveling wave ultrasonic motor using a radial bending mode. *IEEE Trans. Ultrason. Ferroelectr. Freq. Control* **2014**, *61*, 197–202. [CrossRef] [PubMed]
9. Liu, Y.; Chen, W.; Liu, J.; Yang, X. A High-Power Linear Ultrasonic Motor Using Bending Vibration Transducer. *IEEE Trans. Ind. Electron.* **2013**, *60*, 5160–5166. [CrossRef]
10. Liu, Y.; Chen, W.; Yang, X.; Liu, J. A Rotary Piezoelectric Actuator Using the Third and Fourth Bending Vibration Modes. *IEEE Trans. Ind. Electron.* **2014**, *61*, 4366–4373. [CrossRef]
11. Cheng, T.; Fu, X.; Liu, W.; Lu, X.; Chen, X.; Wang, Y.; Bao, G. Airfoil-based cantilevered polyvinylidene fluoride layer generator for translating amplified air-flow energy. *Renew. Energy* **2019**, *135*, 399–407. [CrossRef]
12. Zhao, L.; Yang, Y. An impact-based broadband aeroelastic energy harvester for concurrent wind and base vibration energy harvesting. *Appl. Energy* **2018**, *212*, 233–243. [CrossRef]
13. Wang, J.; Zhou, S.; Zhang, Z.; Yurchenko, D. High-performance piezoelectric wind energy harvester with Y-shaped attachments. *Energy Convers. Manag.* **2019**, *181*, 645–652. [CrossRef]
14. Lai, Z.H.; Wang, J.L.; Zhang, C.L.; Zhang, G.Q.; Yurchenko, D. Harvest wind energy from a vibro-impact DEG embedded into a bluff body. *Energy Convers. Manag.* **2019**, *199*, 111993. [CrossRef]
15. Yang, Z.; Zhou, S.; Zu, J.; Inman, D. High-Performance Piezoelectric Energy Harvesters and Their Applications. *Joule* **2018**, *2*, 642–697. [CrossRef]
16. Gong, Y.; Yang, Z.; Shan, X.; Sun, Y.; Xie, T.; Zi, Y. Capturing Flow Energy from Ocean and Wind. *Energies* **2019**, *12*, 2184. [CrossRef]
17. van Rooij, A.C.L.M.; Nitzsche, J.; Dwight, R.P. Energy budget analysis of aeroelastic limit-cycle oscillations. *J. Fluids Struct.* **2017**, *69*, 174–186. [CrossRef]

18. Zhou, S.; Wang, J. Dual serial vortex-induced energy harvesting system for enhanced energy harvesting. *AIP Adv.* **2018**, *8*, 075221. [CrossRef]
19. Xiang, J.; Wu, Y.; Li, D. Energy harvesting from the discrete gust response of a piezoaeroelastic wing: Modeling and performance evaluation. *J. Sound Vib.* **2015**, *343*, 176–193. [CrossRef]
20. Wu, Y.; Li, D.; Xiang, J.; Da Ronch, A. Piezoaeroelastic energy harvesting based on an airfoil with double plunge degrees of freedom: Modeling and numerical analysis. *J. Fluids Struct.* **2017**, *74*, 111–129. [CrossRef]
21. Yan, Z.; Abdelkefi, A. Nonlinear characterization of concurrent energy harvesting from galloping and base excitations. *Nonlinear Dyn.* **2014**, *77*, 1171–1189. [CrossRef]
22. Yan, Z.; Abdelkefi, A.; Hajj, M.R. Piezoelectric energy harvesting from hybrid vibrations. *Smart Mater. Struct.* **2014**, *23*, 025026. [CrossRef]
23. De Lorenzo, G.; Milewski, J.; Fragiacomo, P. Theoretical and experimental investigation of syngas-fueled molten carbonate fuel cell for assessment of its performance. *Int. J. Hydrogen Energy* **2017**, *42*, 28816–28828. [CrossRef]
24. Farhad, S.; Hamdullahpur, F.; Yoo, Y. Performance evaluation of different configurations of biogas-fuelled SOFC micro-CHP systems for residential applications. *Int. J. Hydrogen Energy* **2010**, *35*, 3758–3768. [CrossRef]
25. Zhang, M.; Hu, G.; Wang, J. Bluff body with built-in piezoelectric cantilever for flow-induced energy harvesting. *Int. J. Energy Res.* **2020**, *44*, 3762–3777. [CrossRef]
26. Kupecki, J.; Skrzypkiewicz, M.; Wierzbicki, M.; Stepien, M. Experimental and numerical analysis of a serial connection of two SOFC stacks in a micro-CHP system fed by biogas. *Int. J. Hydrogen Energy* **2017**, *42*, 3487–3497. [CrossRef]
27. Baldinelli, A.; Barelli, L.; Bidini, G. Upgrading versus reforming: An energy and exergy analysis of two Solid Oxide Fuel Cell-based systems for a convenient biogas-to-electricity conversion. *Energy Convers. Manag.* **2017**, *138*, 360–374. [CrossRef]
28. Bochentyn, B.; Błaszczak, P.; Gazda, M.; Fuerte, A.; Wang, S.F.; Jasiński, P. Investigation of praseodymium and samarium co-doped ceria as an anode catalyst for DIR-SOFC fueled by biogas. *Int. J. Hydrogen Energy* **2020**, *45*, 29131–29142. [CrossRef]
29. De Lorenzo, G.; Fragiacomo, P. Electrical and thermal analysis of an intermediate temperature IIR-SOFC system fed by biogas. *Energy Sci. Eng.* **2018**, *6*, 60–72. [CrossRef]
30. De Lorenzo, G.; Corigliano, O.; Lo Faro, M.; Frontera, P.; Antonucci, P.; Zignani, S.C.; Trocino, S.; Mirandola, F.A.; Aricò, A.S.; Fragiacomo, P. Thermoelectric characterization of an intermediate temperature solid oxide fuel cell system directly fed by dry biogas. *Energy Convers. Manag.* **2016**, *127*, 90–102. [CrossRef]
31. Akaydin, H.D.; Elvin, N.; Andreopoulos, Y. Energy Harvesting from Highly Unsteady Fluid Flows using Piezoelectric Materials. *J. Intell. Mater. Syst. Struct.* **2010**, *21*, 1263–1278. [CrossRef]
32. Goushcha, O.; Akaydin, H.D.; Elvin, N.; Andreopoulos, Y. Energy harvesting prospects in turbulent boundary layers by using piezoelectric transduction. *J. Fluids Struct.* **2015**, *54*, 823–847. [CrossRef]
33. Hu, J.; Porfiri, M.; Peterson, S.D. Energy transfer between a passing vortex ring and a flexible plate in an ideal quiescent fluid. *J. Appl. Phys.* **2015**, *118*, 114902. [CrossRef]
34. Shi, S.; New, T.H.; Liu, Y. Flapping dynamics of a low aspect-ratio energy-harvesting membrane immersed in a square cylinder wake. *Exp. Therm. Fluid Sci.* **2013**, *46*, 151–161. [CrossRef]
35. Dai, H.L.; Abdelkefi, A.; Wang, L. Piezoelectric energy harvesting from concurrent vortex-induced vibrations and base excitations. *Nonlinear Dyn.* **2014**, *77*, 967–981. [CrossRef]
36. Wang, J.; Ran, J.; Zhang, Z. Energy Harvester Based on the Synchronization Phenomenon of a Circular Cylinder. *Math. Probl. Eng.* **2014**, *2014*, 1–9. [CrossRef]
37. Wang, J.; Tang, L.; Zhao, L.; Hu, G.; Song, R.; Xu, K. Equivalent circuit representation of a vortex-induced vibration-based energy harvester using a semi-empirical lumped parameter approach. *Int. J. Energy Res.* **2020**, *44*, 4516–4528. [CrossRef]
38. Zou, Q.; Ding, L.; Wang, H.; Wang, J.; Zhang, L. Two-degree-of-freedom flow-induced vibration of a rotating circular cylinder. *Ocean Eng.* **2019**, *191*, 106505. [CrossRef]
39. Wang, J.; Hu, G.; Su, Z.; Li, G.; Zhao, W.; Tang, L.; Zhao, L. A cross-coupled dual-beam for multi-directional energy harvesting from vortex induced vibrations. *Smart Mater. Struct.* **2019**, *28*, 12LT02. [CrossRef]
40. Gong, Y.; Shan, X.; Luo, X.; Pan, J.; Xie, T.; Yang, Z. Direction-adaptive energy harvesting with a guide wing under flow-induced oscillations. *Energy* **2019**, *187*, 115983. [CrossRef]
41. Abdelkefi, A.; Ghommem, M.; Nuhait, A.O.; Hajj, M.R. Nonlinear analysis and enhancement of wing-based piezoaeroelastic energy harvesters. *J. Sound Vib.* **2014**, *333*, 166–177. [CrossRef]
42. Abdelkefi, A.; Nayfeh, A.H.; Hajj, M.R. Modeling and analysis of piezoaeroelastic energy harvesters. *Nonlinear Dyn.* **2011**, *67*, 925–939. [CrossRef]
43. Dias, J.A.C.; De Marqui, C.; Erturk, A. Three-Degree-of-Freedom Hybrid Piezoelectric-Inductive Aeroelastic Energy Harvester Exploiting a Control Surface. *AIAA J.* **2015**, *53*, 394–404. [CrossRef]
44. Shoele, K.; Mittal, R. Energy harvesting by flow-induced flutter in a simple model of an inverted piezoelectric flag. *J. Fluid Mech.* **2016**, *790*, 582–606. [CrossRef]
45. Wang, J.; Geng, L.; Zhou, S.; Zhang, Z.; Lai, Z.; Yurchenko, D. Design, modeling and experiments of broadband tristable galloping piezoelectric energy harvester. *Acta Mech. Sin.* **2020**. [CrossRef]
46. Yang, K.; Wang, J.; Yurchenko, D. A double-beam piezo-magneto-elastic wind energy harvester for improving the galloping-based energy harvesting. *Appl. Phys. Lett.* **2019**, *115*, 193901. [CrossRef]

47. Wang, J.; Tang, L.; Zhao, L.; Zhang, Z. Efficiency investigation on energy harvesting from airflows in HVAC system based on galloping of isosceles triangle sectioned bluff bodies. *Energy* **2019**, *172*, 1066–1078. [CrossRef]
48. Tan, T.; Hu, X.; Yan, Z.; Zhang, W. Enhanced low-velocity wind energy harvesting from transverse galloping with super capacitor. *Energy* **2019**, *187*, 115915. [CrossRef]
49. Taylor, G.W.; Burns, J.R.; Kammann, S.A.; Powers, W.B.; Welsh, T.R. The Energy Harvesting Eel: A small subsurface ocean/river power generator. *IEEE J. Ocean. Eng.* **2001**, *26*, 539–547. [CrossRef]
50. Allen, J.J.; Smits, A.J. Energy harvesting eel. *J. Fluids Struct.* **2001**, *15*, 629–640. [CrossRef]
51. Akaydın, H.D.; Elvin, N.; Andreopoulos, Y. Wake of a cylinder: A paradigm for energy harvesting with piezoelectric materials. *Exp. Fluids* **2010**, *49*, 291–304. [CrossRef]
52. Weinstein, L.A.; Cacan, M.R.; So, P.M.; Wright, P.K. Vortex shedding induced energy harvesting from piezoelectric materials in heating, ventilation and air conditioning flows. *Smart Mater. Struct.* **2012**, *21*, 045003. [CrossRef]
53. Li, S.; Sun, Z. Harvesting vortex energy in the cylinder wake with a pivoting vane. *Energy* **2015**, *88*, 783–792. [CrossRef]
54. Yu, Y.; Liu, Y. Flapping dynamics of a piezoelectric membrane behind a circular cylinder. *J. Fluids Struct.* **2015**, *55*, 347–363. [CrossRef]
55. Song, R.; Shan, X.; Lv, F.; Li, J.; Xie, T. A Novel Piezoelectric Energy Harvester Using the Macro Fiber Composite Cantilever with a Bicylinder in Water. *Appl. Sci.* **2015**, *5*, 1942. [CrossRef]
56. Li, H.; Sumner, D. Vortex shedding from two finite circular cylinders in a staggered configuration. *J. Fluids Struct.* **2009**, *25*, 479–505. [CrossRef]
57. Abdelkefi, A.; Scanlon, J.M.; McDowell, E.; Hajj, M.R. Performance enhancement of piezoelectric energy harvesters from wake galloping. *Appl. Phys. Lett.* **2013**, *103*, 033903. [CrossRef]
58. Hu, G.; Wang, J.; Su, Z.; Li, G.; Peng, H.; Kwok, K.C.S. Performance evaluation of twin piezoelectric wind energy harvesters under mutual interference. *Appl. Phys. Lett.* **2019**, *115*, 073901–073905. [CrossRef]
59. Shan, X.; Song, R.; Fan, M.; Xie, T. Energy-Harvesting Performances of Two Tandem Piezoelectric Energy Harvesters with Cylinders in Water. *Appl. Sci.* **2016**, *6*, 230. [CrossRef]

 *micromachines*

*Article*

# Data-Driven Optimization of Piezoelectric Energy Harvesters via Pattern Search Algorithm

**Yang Huang** [1,2] , **Zhiran Yi** [1,3] , **Guosheng Hu** [1,2] **and Bin Yang** [1,2,*]

[1]  National Key Laboratory of Science and Technology on Micro/Nano Fabrication, Shanghai Jiao Tong University, Shanghai 200240, China; cedarhuang@sjtu.edu.cn (Y.H.); yizhiran@sjtu.edu.cn (Z.Y.); huguosheng2019@sjtu.edu.cn (G.H.)

[2]  Department of Micro/Nano Electronics, Shanghai Jiao Tong University, Shanghai 200240, China

[3]  State Key Laboratory of Mechanical System and Vibration, School of Mechanical Engineering, Shanghai Jiao Tong University, Shanghai 200240, China

*  Correspondence: binyang@sjtu.edu.cn

**Abstract:** A data-driven optimization strategy based on a generalized pattern search (GPS) algorithm is proposed to automatically optimize piezoelectric energy harvesters (PEHs). As a direct search method, GPS can iteratively solve the derivative-free optimization problem. Taking the finite element method (FEM) as the solver and the GPS algorithm as the optimizer, the automatic interaction between the solver and optimizer ensures optimization with minimum human efforts, saving designers' time and performing a more precise exploration in the parameter space to obtain better results. When employing it for the optimization of PEHs, the optimal length and thickness of PZT were 6.0 mm and 4.6 µm, respectively. Compared with reported high-output PEHs, this optimal structure showed an increase of 371% in output power, an improvement by 1000% in normalized power density, and a reduction of 254% in resonant frequency. Furthermore, Spearman's rank correlation coefficient was calculated for evaluating the correlation among geometric parameters and output performance such as resonant frequency and output power, which provides a data-based perspective on the design and optimization of PEHs.

**Keywords:** piezoelectric; energy harvester; optimization; pattern search; FEM; PZT

check for
**updates**

**Citation:** Huang, Y.; Yi, Z.; Hu, G.; Yang, B. Data-Driven Optimization of Piezoelectric Energy Harvesters via Pattern Search Algorithm. *Micromachines* **2021**, *12*, 561. https://doi.org/10.3390/mi12050561

Academic Editor: Ju-Hyuck Lee

Received: 14 April 2021
Accepted: 13 May 2021
Published: 15 May 2021

**Publisher's Note:** MDPI stays neutral with regard to jurisdictional claims in published maps and institutional affiliations.



## 1. Introduction

With the urgent demand of sustainable power supplies for low-power electronic applications such as wireless sensor network systems [1–4] in Internet of Things, implantable medical devices [5,6] and other devices in some extreme environments [7], energy harvesting from the ambient environments has attracted broad attention and provided potential solutions to the periodical replacement of batteries during the last few decades. Vibration energy is ubiquitous and robust mechanical energy that exists widely, including bridges [8], roads [9], human body [10–12], cars [7], etc. Therefore, a variety of vibration-based piezoelectric energy harvesters (PEHs) with different structures have been proposed and studied [13–15]. Among these, the cantilever structure is widely used due to structure simplicity and the high average strain obtained by a given input force [2,16–19].

For cantilever PEHs, a multi-parameter coupling problem exists for obtaining high-efficiency energy conversion. To improve the output performance of cantilever PEHs, researchers have studied the effects of different geometric parameters on output performance [20,21]. He et al. [22] and Jia et al. [23] proposed that the optimal mass-beam length ratio is 0.6~0.7 within linear response. Hu et al. [18] investigated the optimal length of the piezoelectric layers based on theoretical analysis, FEM simulation and experimental verification, from which they discovered that the optimal length ratio of piezoelectric layers and the beam is approximately 0.2. Furthermore, Hu et al. [18] also reported that the optimal PZT layout length decreases as cantilever width increases. However, investigations on various geometric parameters in the studies above constitute single-variable optimization. Moreover, the geometric parameters were manually set and tested at a fixed interval based on the researchers' experience and intuition, which required the researcher to spend more time in trying each possible combination of geometric parameters to obtain the optimal output performance. To overcome these shortcomings, several data-driven optimization strategies combining different algorithms to maximize the output performance of PEHs have been proposed recently. For example, data-driven optimization can solve the optimization problems that are difficult to formulate or solve, in that they are non-convex, multi-objective or multi-modal; it can solve optimization problems based on derivative-free data. Ghoddus et al. [24] presented an optimization approach based on the Particle swarm optimization (PSO) algorithm to maximize the output power of PEHs with four different structures, which simultaneously optimized multiple geometric parameters. In addition, Nabavi and Zhang [25,26] proposed an analytical model for PEHs with proof mass and used a genetic algorithm to simultaneously optimize multiple objectives, including resonant frequency, output power and device volume. However, finding the optimal solution to complex high-dimensional problems such as the multi-parameter coupling problem for PEHs may require expensive objective function evaluations. Therefore, as population-based stochastic optimization techniques, the PSO and GA algorithm implemented in the previous works [24–26] may encounter low efficiency when dealing with the multi-parameter coupling problems. Moreover, parameter tuning is needed to obtain a better convergence for the PSO algorithm and GA [27], such as particle number, accelerate constant, inertia weight and population size. Although the global search ability of GA and PSO may be better, the generalized pattern search (GPS) algorithm is more suitable for the time-consuming multi-parameter coupling problem due to its effortless parameter tuning and relatively fewer objective function evaluations [28].

In this paper, we proposed a data-driven optimization strategy based on the GPS algorithm. The GPS algorithm is one of the direct search methods, which can be effectively used in derivative-free optimization. By implementing the proposed scheme, multiple geometric parameters ($l_p$, $w$, $t_p$, and $t_b$) were simultaneously optimized for maximum output performance based on the data without building an analytical model. Using the proposed optimization method can not only efficiently optimize the geometric parameters of PEHs to achieve high output performance but also save time for analytical modeling and complex parameter tuning. Furthermore, Spearman's rank correlation coefficient was calculated for evaluating the correlation among geometric parameters and output performance, providing a data-based perspective on the design of PEHs.

## 2. Methods

The unimorph PEH is composed of a piezoelectric layer, a structural layer, and a proof mass. The piezoelectric layer partially covers the flexible structural layer along the constraint end, as shown in Figure 1a. A tungsten proof mass is fixed at the free end of the structural layer, increasing average strain under given excitation and lowering the resonant frequency to meet the low-frequency ambient vibration. Figure 1b illustrates the annotation of this unimorph PEH. The data-driven optimization scheme mainly includes a FEM solver and a GPS optimizer. Figure 1c depicts the overall function of the presented strategy. In this paper, FEM simulation was carried out using COMSOL Multiphysics (Version 5.4) and the material properties used in FEM simulation are listed in Table 1. The GPS algorithm was implemented in MATLAB (Version 2020a). The communications between solver and optimizer were implemented using COMSOL LiveLink™ for the MATLAB module, which facilitated the data-driven optimization strategy. As direct search methods, different types of pattern search algorithms have been utilized for solving engineering optimization problems [28–30]. The convergence analysis of GPS algorithm has been performed by Audet and Dennis [31].

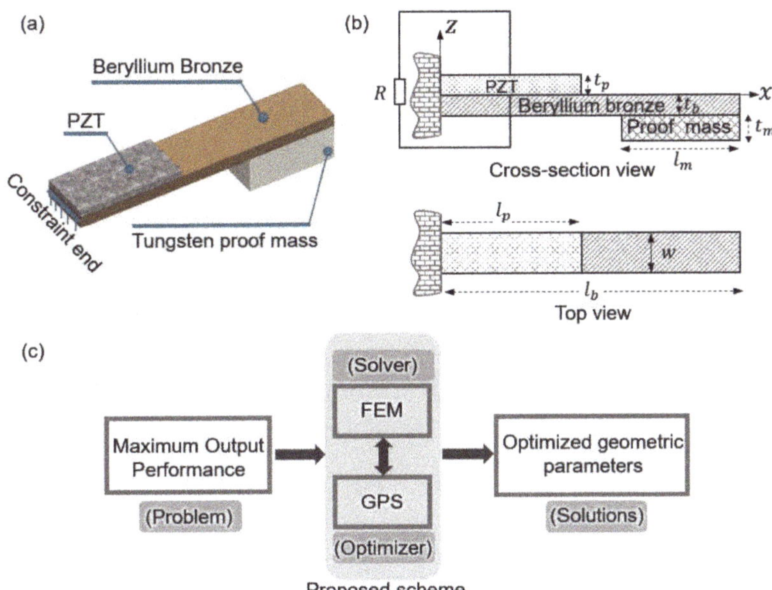



**Figure 1.** Schematic diagram of the piezoelectric energy harvester (PEH) and the proposed strategy. (**a**) Three-dimensional view. (**b**) Annotation for geometric parameters of the cantilever PEH. (**c**) Overview of the presented optimization working process.

**Table 1.** The material properties used in FEM simulation.

| Parameters | Young's Modulus GPa | Density kg/m³ | Poisson Ratio | Elasticity Matrix GPa | Piezoelectric Coupling Matrix C/m² |
|---|---|---|---|---|---|
| PZT | - | 7500 | 0.31 | {127.205, 80.2122, 127.205, 84.6702, 84.6702, 117.436, 0, 0, 0, 22.9885, 0, 0, 0, 0, 22.9885, 0, 0, 0, 0, 0, 23.4742} | {0, 0, −6.62281, 0, 0, −6.62281, 0, 0, 23.2403, 0, 17.0345, 0, 17.0345, 0, 0, 0, 0, 0} |
| Beryllium copper | 128 | 8250 | 0.3 | - | - |
| Tungsten | 411 | 19,350 | 0.28 | - | - |

At each iteration, the GPS searches a set of points around the current point called mesh, finding a better point whose value of the objective function is lower than the value before. Then, the better point is set as the current point at the next iteration. Based on this procedure (Polling), the GPS finds a sequence of points that approaches the optimum, and it does not stop until the convergence is met. Specifically, let $M_k$ denote the mesh at $k$-th iteration, and $x_k^{(i)}$ denote the $i$-th point in the $M_k$. The mesh is defined as Equation (1):

$$M_k \triangleq \left\{ x \in \mathbb{R}^n \mid x = x'_k + \Delta_k \cdot v_k, \; k \in \{1, 2, \cdots, n\} \right\} \tag{1}$$

where $x'_k$ denotes the current point at the $k$-th iteration while $\Delta_k$ denotes mesh size, and $v_k$ denotes pattern. Pattern is a set of vectors used to determine the generation of mesh. Suppose $F(x)$ denotes the objective function in the optimization problem. At each iteration, GPS computes $F\left(x_k^{(i)}\right)$ and looks for a better point $x_k^{(j)}$ so that $F\left(x_k^{(j)}\right) < F(x_k')$; then, $x'_{k+1} = x_k^{(j)}$ and $\Delta_{k+1} = 2 \cdot \Delta_k$ are set. Otherwise, if the polling fails to find a better point at $k^{th}$ iteration, then $x'_{k+1} = x'_k$ and $\Delta_{k+1} = 1/2 \cdot \Delta_k$. The computation in the mesh at each iteration is called polling. In addition, the search method runs before polling can select a different current point, which may accelerate the optimization if tuned well. Various search methods can be set, including the genetic algorithm, Latin hypercube search, etc.

The proposed optimization strategy uses FEM simulation as the solver and GPS algorithm as the optimizer, whose workflow is depicted in Figure 2. The initialization step is mainly used for setting parameters in the GPS algorithm, including initial point, the searching and polling method, stopping criteria and other parameters used to accelerate the optimization. Moreover, a new set of geometric parameters generated by GPS is used for FEM modelling, and then, the FEM simulation is performed in order to extract the solutions for feedback to the GPS algorithm. In order to efficiently harvest energy in daily life such as cars, bridges and the human body, the development of PEHs tends to have high output power density and low resonant frequency. Therefore, the optimization objective function here is defined as normalized power density or output power. The definition of normalized power density is shown as Equation (2), which is a function of various geometric parameters and external excitation:

$$P_n = \frac{P}{a^2 \cdot f_1 \cdot V_{eff}} \tag{2}$$

where $P_n$ denotes the normalized power density, $P$ denotes the output power, $f_1$ denotes the first-modal resonant frequency, $V_{eff}$ denotes the effective volume of the specific structure, and $a$ represent the excitation acceleration. The maximizing of normalized power density may indicate the trend of maximizing output power and minimizing the first-modal resonant frequency and effective volume. Therefore, the optimization problem is defined as below:

$$
\begin{aligned}
&maximize && P_n \ or \ P \\
&subject\ to && 0.5\text{mm} \le l_p \le 15\,\text{mm} \\
& && 0.5\,\text{mm} \le w \le 3\,\text{mm} \\
& && 0.001\,\text{mm} \le t_p \le 0.05\,\text{mm} \\
& && 0.03\,\text{mm} \le t_b \le 0.05\,\text{mm}
\end{aligned} \tag{3}
$$

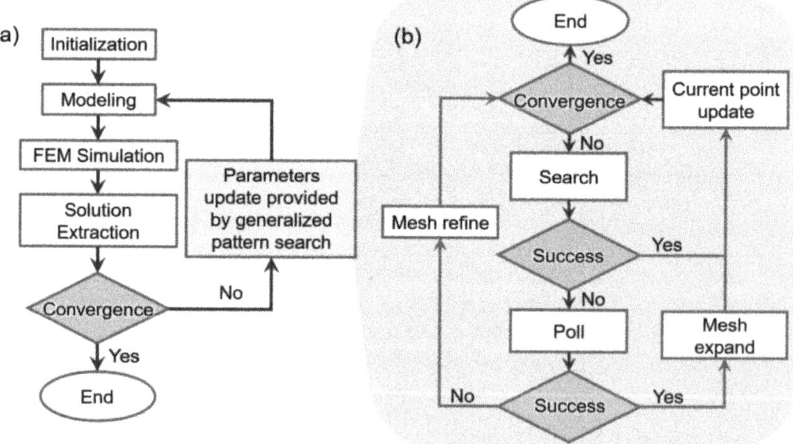


**Figure 2.** Workflow of the proposed data-driven optimization strategy. (**a**) Overall working mechanism. (**b**) Detailed workflow of generalized pattern search.

## 3. Results and Discussion

Utilizing the well-described micromachining processes and experimental setup in our previous work [17], we first fabricated and tested four devices with different length PZT layers to validate the effectiveness of the proposed data-driven optimization strategy, and then, individually optimized the PZT length and the proof mass length intending to maximize output power and compared the optimized result with the previous works [18,23]. The experimental setup is shown in Figure 3a. Controlled by the vibration controller through the power amplifier (YE2706A, Sinocera Piezotronics, Inc., Yangzhou, China), the force applied on different PEHs is generated by a shaker (JK-2, Sinocera Piezotronics, Inc.) and is monitored by a force sensor (208C02, PCB Piezotronics, New York, NY, USA).

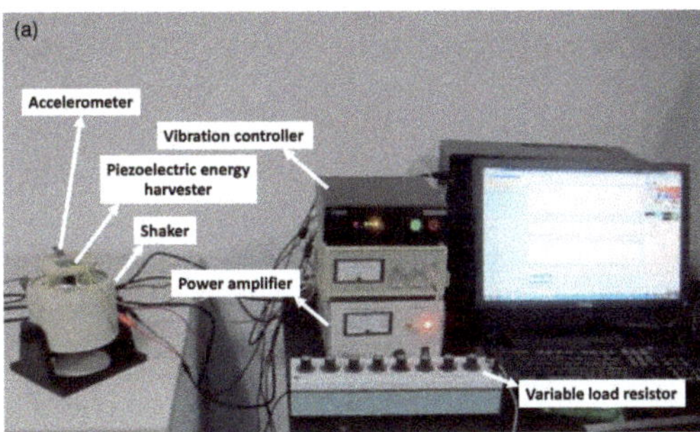

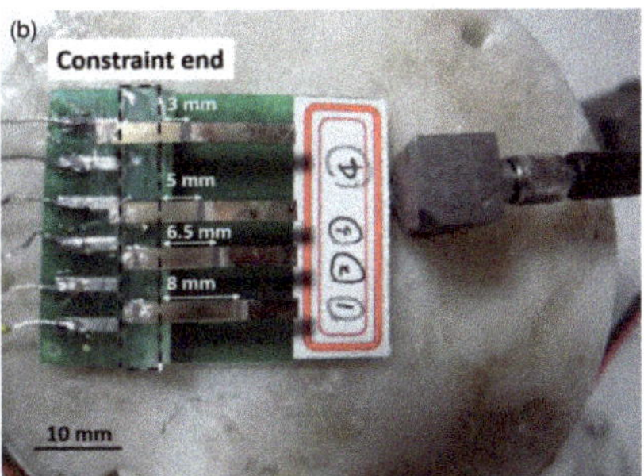


**Figure 3.** (**a**) Photograph of the experimental setup. (**b**) Photograph of the fabricated harvesters with the varied PZT layer length.

Firstly, the bulk PZT (300 μm) and beryllium bronze (50 μm) were polished so as to increase the bonding strength. Then, the bulk PZT was bonded on the beryllium bronze using conductive silver epoxy in a vacuum oven at 175 °C for 3.5 h.

Next, the bulk PZT was thinned to around 50 μm through chemical mechanical thinning and polishing. After that, 20/200 nm Cr/Au was sputtered on the polished surface of the bulk PZT thick film as an electrode. Finally, the cantilevered PEHs were patterned using the ultraviolet laser method, and the proof mass was assembled at the free end. The fabricated harvesters with different PZT layer length are illustrated in Figure 3b. Under 1.0 g acceleration, the PEHs were connected in serial with the external resistance, which varied from 1 kΩ to 1000 kΩ to determine the optimum load resistance and output power. For performing FEM simulation, a geometrical configuration was considered where the free end was mounted with a proof mass, and the upper surface of the PZT was grounded. In addition, chamfer was added between the proof mass and structural layer to avoid stress singularities at the reentrant corners. Meshes were created according to the shape of the geometry. Then, the detailed meshes were determined by carrying out a mesh convergence study. Figure 4a presents the meshes of the PEHs constructed in COMSOL software. Here, we used skewness, the default quality measure in COMSOL software, to evaluate the mesh quality, which is a suitable measure for most types of meshes. To extract the optimum output power of each PEH, the FEM simulation firstly ran an eigenfrequency study to obtain the first-modal eigenfrequency of the PEHs, and then ran a frequency domain study under the first-modal eigenfrequency with an auxiliary sweep of different external resistance. The excitation was set as body load with an acceleration of 1.0 g (9.8 m/s$^2$). Using the proposed scheme, the procedures mentioned above will not stop until the GPS algorithm meets convergence, and the structure of PEHs will keep updated to seek the optimum geometric parameters.

Figure 4b shows a comparison of normalized output power for the presented PEHs with the varied PZT layer length under different external resistance between FEM simulation and experimental results. Compared with the experimental data, the FEM simulation showed agreement in the trend of maximum output power with varied PZT layer length, which proved the validation of the FEM simulation in determining the optimal geometric parameters. Especially when the length of the PZT layer is 3.0 mm, the FEM simulation fits well with the experimental data. The difference in output power corresponding to external resistance between FEM simulation and experimental results may come from manufacturing and testing errors. As the length of the PZT layer increases, the manufacturing errors accumulate. The difference between FEM and the experiment decreases as the length of PZT decreases. Moreover, the mechanical damping (set as 0.015) was assumed to be a constant in the FEM simulation, which may also lead to errors in power prediction for different structure [32]. Then, two single-variable optimizations (PZT length and mass length) for maximum output performance were performed and compared with the previous works [18,23]. For the cantilever harvester with proof mass, Hu et al. [18] and Jia et al. [23] have, respectively, optimized the PZT length and the mass length to maximize output power, from which they have concluded that the optimal ratio of PZT length and total length is approximately 0.2 (3 mm/15 mm), while the optimal mass-beam length ratio is 0.6~0.7.

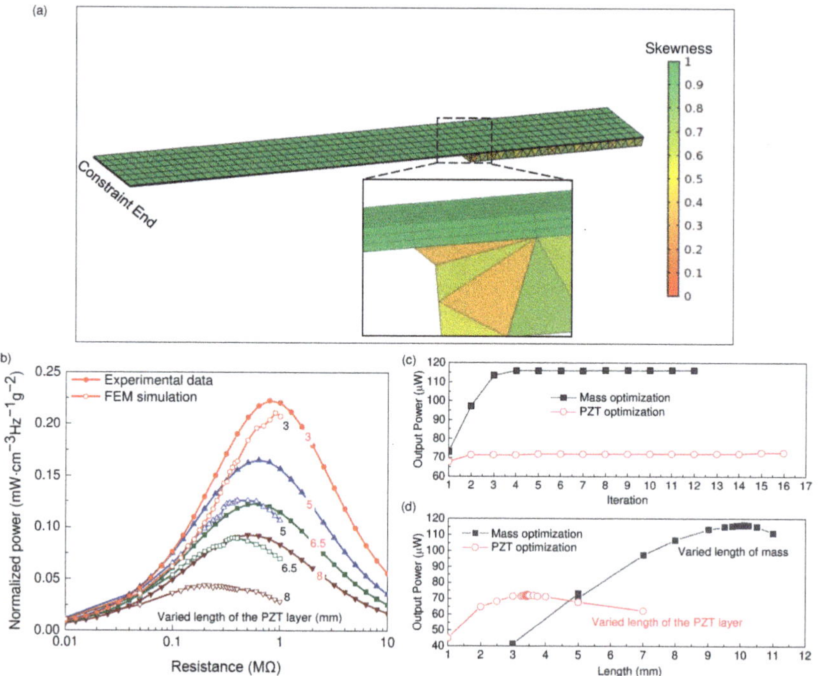

**Figure 4.** (**a**) The mesh of the PEHs in FEM simulation. Validation of the proposed scheme by (**b**) comparison of normalized output power for the presented PEHs with the varied PZT layer length under different external resistance between FEM simulation and experimental results. (**c**) Convergence curve of output power for each single-variable optimization given by the proposed scheme. (**d**) All results searched by the proposed scheme for each single-variable optimization. The applied acceleration amplitude is 1.0 g.

Figure 4c shows the convergence curve of mass and PZT length optimization. It can be observed that the GPS algorithm can efficiently optimize the proposed optimization problems within two and four iterations, respectively, for PZT length and mass length optimization. Figure 4d presents all results evaluated by the GPS algorithm, which demonstrates the trend of output power versus varied length of the mass and PZT. The optimal PZT length and mass length are 3.437 mm and 10.090 mm corresponding to the ratio of the total length of 0.229 and 0.672, respectively. The agreement of the optimized results and that of the previous works [18,23] proves the efficiency and effectiveness of the proposed data-driven optimization method and the FEM model. Furthermore, unlike the optimization approaches in the previous works [18,23], by which researchers manually varied the geometric parameters at fixed intervals to explore the parameter space for maximum output performance, the proposed method automatically performs the optimization task once set, saving researchers time compared to trial-and-error approaches. In addition, a more precise exploration in the parameter space can be carried out by the proposed scheme compared with the manual trial-and-error approaches. Respectively labeled as 1OPT-1 and 1OPT-2, the geometric parameters and output performance of the above two single-variable optimizations are summarized in Table 2.

**Table 2.** Geometric parameters and results of our calculations.

| Parameters | $l_p$ mm | $w$ mm | $l_m$ mm | $t_p$ μm | $t_b$ μm | $f_1$ Hz | $P$ μW | $P_D$ mW·cm$^{-3}$ | $P_n$ mW·cm$^{-3}$·Hz$^{-1}$·g$^{-2}$ | Runtime min |
|---|---|---|---|---|---|---|---|---|---|---|
| Ref [18] | 3.000 | 2.50 | 5.00 | 50.00 | 50.00 | 66.96 | 70.28 | 14.80 | 0.221 | - |
| 1OPT-1 | 3.437 | 2.50 | 5.00 | 50.00 | 50.00 | 70.19 | 72.70 | 15.13 | 0.216 | 12 |
| 1OPT-2 | 3.000 | 2.50 | 10.09 | 50.00 | 50.00 | 81.46 | 116.17 | 15.92 | 0.195 | 13 |
| 4OPT-1 | 6.001 | 3.00 | 5.00 | 4.625 | 30.00 | 26.38 | 261.02 | 58.87 | 2.232 | 61 |

$l_b$ = 15.00 mm; $t_m$ = 0.20 mm; Acceleration = 1 g.

After verifying the effectiveness of the proposed scheme, further optimization for the unimorph PEHs was implemented to obtain maximum output power and minimum resonant frequency. The optimization problem is depicted as Equation (3). The geometric parameters to be optimized were set as $l_p$, $w$, $t_p$ and $t_b$. As the thickness of the whole beam becomes thinner, the higher the probability of device failure. Based on the experimental data in the previous work [18], the upper and lower bound of $t_b$ was set as 30–50 mm from a conservative consideration. The optimal results and the initial point in the optimization, labeled as 4OPT-1 and Ref, are summarized in Table 2.

Figure 5a presents the trajectory of the geometric parameters and output power during optimization, which directly shows how the GPS algorithm optimizes the defined optimizable parameters to obtain better values of objectives function. Each point, respectively, represents the best point at every iteration. As shown in Figure 5a and Table 2, the optimal length and thickness of PZT are 6.0 mm and 4.6 μm, respectively. In addition, the reduction in thickness of the structural layer also contributes to the improvement of output power. Figure 5b demonstrates the comparison of the output performance between the reference structure (Ref) and optimized structure (4OPT-1). It can be observed that the resonant frequency of 4OPT-1 is 2.54 times lower than that of Ref, while the output power and normalized power density were, respectively, about 3.71 and 10.10 times larger, showing a substantial improvement in output performance. Numerically, the reduction in resonant frequency greatly contributed to the improvement of normalized power density according to the definition given by Equation (2). By calculating the second derivative of the tip displacement at the first-modal vibration, the strain of the piezoelectric layer along the arc length ($S_1(x, y, t) = -y\frac{\partial^2 z(x,t)}{\partial x^2}$) is presented in Figure 5c, which demonstrates a higher average strain distribution of 4OPT-1 than that of Ref, leading to an improvement in the output performance of 4OPT-1. However, it is noted that the effects of a reduction in material properties such as piezoelectricity with the decrease in the thickness of PZT are neglected theoretically. Although the output performance of 4OPT-1 may be augmented without the consideration of the effects of material properties, the optimal result (4OPT-1) given by the proposed data-driven optimization strategy provides the trend of improving output performance by tuning geometric parameters.

Furthermore, a comparison between the GPS algorithm and GA for the four-variable optimization problem has been implemented and is depicted in Figure 6, which shows the convergence curve versus the number of function evaluations using the GPS algorithm and GA with different population size. The parameters used in GA are the default settings in MATLAB; that is, the mutation method and crossover method are constraint dependent, and the selection method is stochastically uniform. Each configuration was limited to run approximately 250 times for comparison.

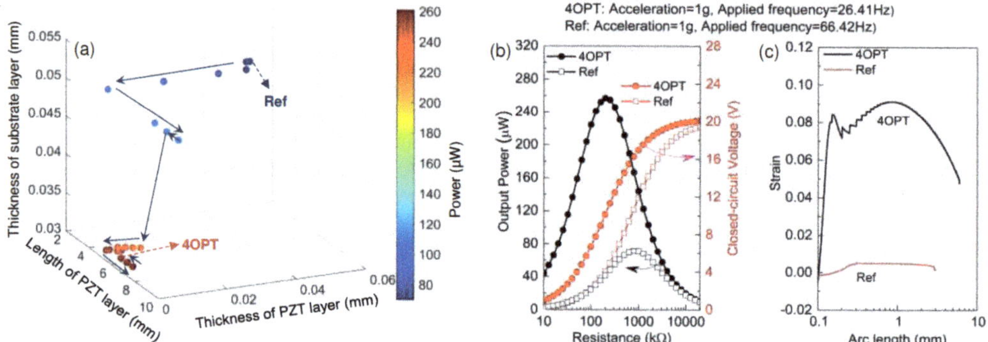

**Figure 5.** Evaluation of the proposed data-driven optimization strategy from various perspectives. (**a**) Trajectory of the parameters during optimization. (**b**) Output power versus external resistance for previous reported high-output PEH and optimized structure. (**c**) Strain distributions along the arc length of previous reported high-output PEH and optimized structure.

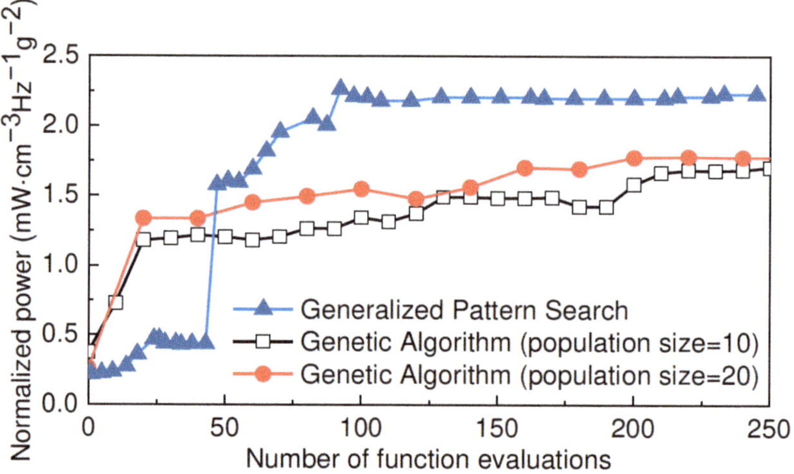

**Figure 6.** Comparison among the efficiency of generalized pattern search algorithm and genetic algorithm with different population size.

To some extent, the number of function evaluations can reflect the running time due to having the same solver (FEM simulation) for each configuration. Although the genetic algorithm can obtain a higher normalized output power at the beginning, the GPS algorithm succeeds in achieving better performance than the GA algorithm with a population size of 10 or 20 after evaluation 250 times. Furthermore, the GPS algorithm requires fewer function evaluation than the GA. The reasons why the pattern search algorithm is more efficient than the GA algorithm with a population size of 10 or 20 are numerous, including the characteristics of the optimization problem, the complex parameter tuning of GA algorithm, and the size of parameter space. Theoretically, the performance of GA may be improved by increasing the population size, whereas the running time may increase for the proposed optimization problem in this work, possibly leading to a low efficiency of optimization.

Figure 7a shows a scatterplot among the geometric parameters $l_p$, $w$, $t_p$, $t_b$, resonant frequency and output power. It is observed that the data are not completely randomly distributed in the parameter space due to the minimum searching characteristic of the GPS algorithm and GA. The peak in each histogram in the first four rows indicates the preferable geometric parameters given by the GPS algorithm and GA that may generate better output performance. The contour of the scatterplot of the power and thickness of PZT, the power and thickness of the structural layers, and the power and resonant frequency may present non-linearity among the aforementioned variables. Based on the data in the scatterplot, Spearman's rank correlation coefficient was calculated using Equation (4) to evaluate the strength and direction of the monotonic relationship among the geometric parameters, resonant frequency and output power.

$$r_s = 1 - \frac{6 \sum d_i^2}{n(n^2 - 1)} \tag{4}$$

where $d_i = rg(X_i) - rg(Y_i)$, representing the difference in rank between $X_i$ and $Y_i$; $n$ is the number of observations; $X_i$ and $Y_i$ are variables to be evaluated.

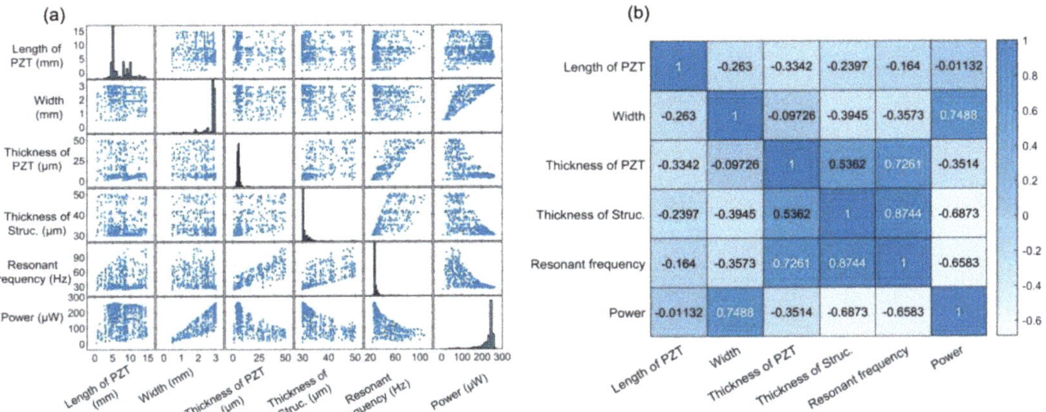

**Figure 7.** (a) Scatterplot and (b) Spearman's rank correlation matrix of the geometric parameters and output performance (resonant frequency and output power).

The correlation matrix with normalized values is presented in Figure 7b. The greater the absolute value of the correlation coefficient, the stronger the monotonic relationship between the evaluated variables. After filtering the identical data, the amount of data available for calculating the Spearman's correlation coefficient is 4506. The maximum value in the correlation matrix is 0.8744 and the corresponding $p$-value under t-distribution is 0, suggesting a strong positive correlation between the thickness of the structural layers and resonant frequency, which means that the designers may need to reduce the thickness of the structural layers to lower the resonant frequency. The $p$-value under t-distribution is calculated using Equation (5):

$$t = r_s \sqrt{\frac{n - 2}{1 - r_s^2}} \tag{5}$$

In addition, the coefficient between power and width is 0.7488, suggesting that the designers may increase the width to improve the output power. However, it is noted that the coefficient of resonant frequency and width is $-0.3573$, indicating a weak negative correlation between them. Expanding the width may increase the resonant frequency. Designers may need to balance the impact of increase on width. Moreover, although the coefficient of power and length of PZT is close to 0, its $p$-value (0.4475) is larger than the general threshold of 0.05, which indicates that this correlation may fail to reject the null hypothesis. As shown in Figure 7a, although the scatterplot of the power and length of PZT presented no monotonic relationship, the upper contour in the scatterplot showed a non-linearity relationship between them, in which the optimal length of PZT is 6.0 mm when the thickness of PZT is 4.6 µm. In addition, in the scatterplot of the power and the length of PZT, although the length of PZT is set as the optimum (6.0 mm), the output power may encounter poor performance without proper setup of other geometric parameters. Therefore, it is necessary to simultaneously optimize multiple parameters for improving the performance of PEHs using the proposed data-driven optimization strategy.

### 4. Conclusions

In summary, a data-driven optimization strategy based on FEM simulation (solver) and a GPS algorithm (optimizer) was proposed and implemented, which can not only optimize the geometric parameters of PEHs to achieve high performance but also save time for analytical modeling and complex parameter tuning. In addition, the proposed strategy can search the variable space more precisely than manually varied geometric parameters at fixed intervals. The effectiveness and efficiency of the proposed scheme were validated by comparing the optimization results with previous works [18,23] and experimental results. Then, a four-variable ($l_p$, $w$, $t_p$ and $t_b$) optimization was further investigated with the aim of maximum output performance. The optimal ratio of the length of PZT and structural layer increases as the thickness of PZT decreases compared with previous works. The optimization results showed that the optimal length ratio is 0.40 (6.0 mm/15.0 mm) when the thickness of PZT is 4.6 µm, while the optimal length ratio is 0.23 (3.4 mm/15.0 mm) when the thickness of PZT is 50 µm. Furthermore, Spearman's rank correlation coefficient was calculated for evaluating the correlation among geometric parameters and output performance such as resonant frequency and output power, providing a data-based perspective on the design and optimization of piezoelectric energy harvesters.

Our evaluation showed that the GPS algorithm exhibited better performance than GA in terms of efficiency, requiring fewer function evaluations than GA. Solving the multi-parameter coupling problem such as that of PEHs may require expensive computation cost. Therefore, utilizing a GPS algorithm can shorten the total optimization time for complex high-dimensional optimization problems in real life. In addition, owing to the extensive application of the FEM simulation, the proposed strategy is not limited to the optimization of PEHs, but can also be used for the optimization of other structures that are challenging to formulate, facilitating the employment of the proposed strategy in other fields.

**Author Contributions:** Conceptualization, Y.H.; methodology, Y.H.; software, Y.H.; validation, Y.H., Z.Y.; formal analysis, Y.H., Z.Y., G.H.; investigation, Y.H., G.H.; resources, B.Y.; data curation, Y.H.; writing—original draft preparation, Y.H.; writing—review and editing, Y.H., Z.Y., G.H., B.Y.; visualization, Y.H., Z.Y.; supervision, B.Y.; project administration, B.Y.; funding acquisition, B.Y. All authors have read and agreed to the published version of the manuscript.

**Funding:** This work is funded by the Pre-research Foundation of Equipment (No. 6142601180202) and the Joint Foundation of Pre-research of Equipment and the Ministry of Education (No. 6141A02022637).

**Data Availability Statement:** Data are contained within the article.

**Acknowledgments:** The authors are also grateful to the Center for Advanced Electronic Materials and Devices (AEMD), Shanghai Jiao Tong University.

**Conflicts of Interest:** The authors declare no conflict of interest.

## References

1. Jung, H.J.; Song, Y.; Hong, S.K.; Yang, C.H.; Hwang, S.J.; Jeong, S.Y.; Sung, T.H. Design and Optimization of Piezoelectric Impact-Based Micro Wind Energy Harvester for Wireless Sensor Network. *Sens. Actuators A Phys.* **2015**, *222*, 314–321. [CrossRef]
2. Liu, H.; Zhong, J.; Lee, C.; Lee, S.-W.; Lin, L. A Comprehensive Review on Piezoelectric Energy Harvesting Technology: Materials, Mechanisms, and Applications. *Appl. Phys. Rev.* **2018**, *5*, 041306. [CrossRef]

3. Dong, C.; Li, S.; Han, R.; He, Q.; Li, X.; Xu, D. Self-Powered Wireless Sensor Network Using Event-Triggered Energy Harvesters for Monitoring and Identifying Intrusion Activities. *IET Power Electron.* **2019**, *12*, 2079–2085. [CrossRef]

4. Fu, H.; Sharif Khodaei, Z.; Aliabadi, M.H.F. An Event-Triggered Energy-Efficient Wireless Structural Health Monitoring System for Impact Detection in Composite Airframes. *IEEE Internet Things J.* **2019**, *6*, 1183–1192. [CrossRef]

5. Yi, Z.; Xie, F.; Tian, Y.; Li, N.; Dong, X.; Ma, Y.; Huang, Y.; Hu, Y.; Xu, X.; Qu, D.; et al. A Battery- and Leadless Heart-Worn Pacemaker Strategy. *Adv. Funct. Mater.* **2020**, *30*, 2000477. [CrossRef]

6. Li, N.; Yi, Z.; Ma, Y.; Xie, F.; Huang, Y.; Tian, Y.; Dong, X.; Liu, Y.; Shao, X.; Li, Y.; et al. Direct Powering a Real Cardiac Pacemaker by Natural Energy of a Heartbeat. *ACS Nano* **2019**, *13*, 2822–2830. [CrossRef]

7. Yi, Z.; Yang, B.; Zhang, W.; Wu, Y.; Liu, J.-Q. Batteryless Tire Pressure Real-Time Monitoring System Driven by an Ultralow-Frequency Piezoelectric Rotational Energy Harvester. *IEEE Trans. Ind. Electron.* **2020**, *68*, 3192–3201. [CrossRef]

8. Kim, S.-H.; Ahn, J.-H.; Chung, H.-M.; Kang, H.-W. Analysis of Piezoelectric Effects on Various Loading Conditions for Energy Harvesting in a Bridge System. *Sens. Actuators A Phys.* **2011**, *167*, 468–483. [CrossRef]

9. Ye, G.; Yan, J.; Wong, Z.J.; Soga, K.; Seshia, A. Optimisation of a Piezoelectric System for Energy Harvesting from Traffic Vibrations. In Proceedings of the 2009 IEEE International Ultrasonics Symposium, Rome, Italy, 20–23 September 2009; pp. 759–762.

10. Pillatsch, P.; Yeatman, E.M.; Holmes, A.S. A Piezoelectric Frequency Up-Converting Energy Harvester with Rotating Proof Mass for Human Body Applications. *Sens. Actuators A Phys.* **2014**, *206*, 178–185. [CrossRef]

11. Yang, J.-H.; Cho, H.-S.; Park, S.-H.; Song, S.-H.; Yun, K.-S.; Lee, J.H. Effect of Garment Design on Piezoelectricity Harvesting from Joint Movement. *Smart Mater. Struct.* **2016**, *25*, 035012. [CrossRef]

12. Wang, Q.; Kim, K.-B.; Woo, S.B.; Song, Y.S.; Sung, T.H. A Flexible Piezoelectric Energy Harvester-Based Single-Layer WS2 Nanometer 2D Material for Self-Powered Sensors. *Energies* **2021**, *14*, 2097. [CrossRef]

13. Zou, H.-X.; Zhao, L.-C.; Gao, Q.-H.; Zuo, L.; Liu, F.-R.; Tan, T.; Wei, K.-X.; Zhang, W.-M. Mechanical Modulations for Enhancing Energy Harvesting: Principles, Methods and Applications. *Appl. Energy* **2019**, *255*, 113871. [CrossRef]

14. Wang, J.; Zhou, S.; Zhang, Z.; Yurchenko, D. High-Performance Piezoelectric Wind Energy Harvester with Y-Shaped Attachments. *Energy Convers. Manag.* **2019**, *181*, 645–652. [CrossRef]

15. Hu, G.; Wang, J.; Su, Z.; Li, G.; Peng, H.; Kwok, K.C.S. Performance Evaluation of Twin Piezoelectric Wind Energy Harvesters under Mutual Interference. *Appl. Phys. Lett.* **2019**, *115*, 073901. [CrossRef]

16. Toprak, A.; Tigli, O. Piezoelectric Energy Harvesting: State-of-the-Art and Challenges. *Appl. Phys. Rev.* **2014**, *1*, 031104. [CrossRef]

17. Yi, Z.; Yang, B.; Li, G.; Liu, J.; Chen, X.; Wang, X.; Yang, C. High Performance Bimorph Piezoelectric MEMS Harvester via Bulk PZT Thick Films on Thin Beryllium-Bronze Substrate. *Appl. Phys. Lett.* **2017**, *111*, 013902. [CrossRef]

18. Hu, Y.; Yi, Z.; Dong, X.; Mou, F.; Tian, Y.; Yang, Q.; Yang, B.; Liu, J. High Power Density Energy Harvester with Non-Uniform Cantilever Structure Due to High Average Strain Distribution. *Energy* **2019**, *169*, 294–304. [CrossRef]

19. Hu, Y.; Wang, R.; Wen, J.; Liu, J.-Q. A Low-Frequency Structure-Control-Type Inertial Actuator Using Miniaturized Bimorph Piezoelectric Vibrators. *IEEE Trans. Ind. Electron.* **2019**, *66*, 6179–6188. [CrossRef]

20. Nisanth, A.; Suja, K.J.; Seena, V. Design and Optimization of MEMS Piezoelectric Energy Harvester for Low Frequency Applications. *Microsyst. Technol.* **2021**, *27*, 251–261. [CrossRef]

21. Mangaiyarkarasi, P.; Lakshmi, P.; Sasrika, V. Enhancement of Vibration Based Piezoelectric Energy Harvester Using Hybrid Optimization Techniques. *Microsyst. Technol.* **2019**, *25*, 3791–3800. [CrossRef]

22. He, X.; Shang, Z.; Cheng, Y.; Zhu, Y. A Micromachined Low-Frequency Piezoelectric Harvester for Vibration and Wind Energy Scavenging. *J. Micromech. Microeng.* **2013**, *23*, 125009. [CrossRef]

23. Jia, Y.; Seshia, A.A. Power Optimization by Mass Tuning for MEMS Piezoelectric Cantilever Vibration Energy Harvesting. *J. Microelectromech. Syst.* **2016**, *25*, 108–117. [CrossRef]

24. Ghoddus, H.; Kordrostami, Z. Harvesting the Ultimate Electrical Power From MEMS Piezoelectric Vibration Energy Harvesters: An Optimization Approach. *IEEE Sens. J.* **2018**, *18*, 8667–8675. [CrossRef]

25. Nabavi, S.; Zhang, L. Design and Optimization of Piezoelectric MEMS Vibration Energy Harvesters Based on Genetic Algorithm. *IEEE Sens. J.* **2017**, *17*, 7372–7382. [CrossRef]

26. Nabavi, S.; Zhang, L. Frequency Tuning and Efficiency Improvement of Piezoelectric MEMS Vibration Energy Harvesters. *J. Microelectromech. Syst.* **2019**, *28*, 77–87. [CrossRef]

27. He, Y.; Ma, W.J.; Zhang, J.P. The Parameters Selection of PSO Algorithm Influencing On Performance of Fault Diagnosis. *MATEC Web Conf.* **2016**, *63*, 02019. [CrossRef]

28. Güneş, F.; Tokan, F. Pattern Search Optimization with Applications on Synthesis of Linear Antenna Arrays. *Expert Syst. Appl.* **2010**, *37*, 4698–4705. [CrossRef]

29. Zhao, Z.; Meza, J.C.; Hove, M.V. Using Pattern Search Methods for Surface Structure Determination of Nanomaterials. *J. Phys. Condens. Matter* **2006**, *18*, 8693–8706. [CrossRef]

30. Abramson, M.A. Mixed Variable Optimization of a Load-Bearing Thermal Insulation System Using a Filter Pattern Search Algorithm. *Optim. Eng.* **2004**, *5*, 157–177. [CrossRef]

31. Audet, C.; Dennis, J.E., Jr. Analysis of Generalized Pattern Searches. *SIAM J. Optim.* **2002**, *13*, 889–903. [CrossRef]

32. Foong, F.M.; Thein, C.K.; Yurchenko, D. On Mechanical Damping of Cantilever Beam-Based Electromagnetic Resonators. *Mech. Syst. Signal Process.* **2019**, *119*, 120–137. [CrossRef]

*Article*

# An Energy Harvester with Temperature Threshold Triggered Cycling Generation for Thermal Event Autonomous Monitoring

Ruofeng Han [1,2], Nianying Wang [1,2,3], Qisheng He [1,2], Jiachou Wang [1,2] and Xinxin Li [1,2,*]

1 State Key Laboratory of Transducer Technology, Shanghai Institute of Microsystem and Information Technology, Chinese Academy of Sciences, Shanghai 200050, China; hanruofeng@mail.sim.ac.cn (R.H.); wangny@mail.sim.ac.cn (N.W.); heqisheng2018@168.com (Q.H.); jiatao-wang@mail.sim.ac.cn (J.W.)
2 School of Microelectronics, University of Chinese Academy of Sciences, Beijing 100049, China
3 School of Information Science and Technology, ShanghaiTech University, Shanghai 201210, China
* Correspondence: xxli@mail.sim.ac.cn; Tel.: +86-21-6213-1794

**Abstract:** This paper proposes a temperature threshold triggered energy harvester for potential application of heat-event monitoring. The proposed structure comprises an electricity generation cantilever and a bimetallic cantilever that magnetically attract together. When the structure is heated to a pre-set temperature threshold, the heat absorption induced bimetallic effect of the bimetallic cantilever will cause sufficient bending of the generation cantilever to get rid of the magnetic attraction. The action triggers the freed generation cantilever into resonance to piezoelectrically generate electricity, and the heated bimetallic cantilever dissipates heat to the environment. With the heat dissipated, the bimetallic cantilever will be restored to attract with the generation cantilever again and the structure returns to the original state. Under continual heating, the temperature threshold triggered cycle is repeated to intermittently generate electric power. In this paper, the temperature threshold of the harvester is modeled, and the harvester prototype is fabricated and tested. The test results indicate that, with the temperature threshold of 71 °C, the harvesting prototype is tested to generate 1.14 V peak-to-peak voltage and 1.077 μW instantaneous power within one cycle. The thermal harvesting scheme shows application potential in heat event-driven autonomous monitoring.

**Keywords:** energy harvester; temperature threshold; piezoelectricity; vibrational cantilever; bimetallic effect

**Citation:** Han, R.; Wang, N.; He, Q.; Wang, J.; Li, X. An Energy Harvester with Temperature Threshold Triggered Cycling Generation for Thermal Event Autonomous Monitoring. *Micromachines* **2021**, *12*, 425. https://doi.org/10.3390/mi12040425

Academic Editor: Qiongfeng Shi and Huicong Liu

Received: 12 March 2021
Accepted: 10 April 2021
Published: 13 April 2021

**Publisher's Note:** MDPI stays neutral with regard to jurisdictional claims in published maps and institutional affiliations.



## 1. Introduction

Wireless sensor networks (WSN) need a huge number of sensing nodes for unattended monitoring to undesired events like industrial equipment failure, human health problem, and environmental disaster, etc. [1]. Many sensor nodes working in the field suffer the common problem of insufficient on-site power supply. For example, in unattended application of forest fire monitoring, field power supply including battery replacement is really difficult. By using on-site energy harvesters, the generated electric power from environment is still often too weak to sustain the whole sensor-node microsystem which includes the sensor, the circuits for signal processing and event identification and the alarming device, et al. Inspired by the event driven analogue-to-information (A-to-I) concept [2], an energy-to-information (E-to-I) idea has arisen in our minds, where the event information of forest firing is obtained simultaneously with the electric-energy converted from the fire induced heat. Among the environmental energies such as solar energy [3], vibration energy [4] and thermal energy [5], thermal energy is one of the richest energy resources that can be converted into electricity.

The common converting mechanism from thermal energy includes thermoelectricity, pyroelectricity, and thermomagneticity [6]. Thermoelectricity is a conversion mechanism based on the Seebeck effect; that is, when two dissimilar electrical conductors or semiconductors are joined together, they make a thermocouple and if the temperature difference is maintained between the two joining junctions, an electromotive force is developed [7,8]. Hao et al. proposed a high efficiency thermoelectric energy harvester with the working temperature between 100 °C and 300 °C by suppressing intrinsic excitation in p-type $Bi_2Te_3$-based materials [9]. Pyroelectric conversion originates from interaction between polarization and temperature change in some dielectric materials [10]. Leng et al. designed a pyroelectric generator based on a polyvinylidene fluoride film to harvest the heat energy from hot/cold water [11]. The device achieves practical application by simply alternating contact with hot flow and cold flow. Thermomagnetic conversion mechanism relies on the effect of heat on the magnetic properties [12]. Chun et al. reported a thermomagneto-electric generator array that is composed of flexible polyvinylidene difluoride bimorph cantilevers [13]. Under thermal gradient, the ferromagnetic phase transition of soft magnet generates vibration and piezoelectric power. Deepak et al. developed a novel hybrid thermomagnetic oscillator by using $(MnNiSi)_{0.7}(Fe2Ge)_{0.3}$ as thermomagnetic material for cooling of the heat load as well as electricity harvesting [14]. The proposed thermomagnetic oscillator can cool the heat load by mechanical oscillation between the load and thermal sink by up to 70 °C. Song et al. developed a magneto-thermal generator by using $La_{0.85}Sr_{0.15}MnO_3$ and $(Ni_{0.6}Cu_{0.2}Zn_{0.2})Fe_2O_4$ composite material. The fabricated magneto-thermal generator shows 0.2 Hz operation frequency and generated maximum outputs of 17 μW under the thermal gradient of 80 °C [15]. Waske et al. developed a thermomagnetic generating concept to convert low-temperature waste heat into electricity. Through a combination of experiment and simulation, it is shown that the pretzel-like topology results in a sign reversal of the magnetic flux, which makes the output voltage and power of the proposed generator two orders of magnitude larger than those in conventional set-ups [16].

In contrast to the above-mentioned methods, we herein propose a thermal energy harvester with a thermo-mechanical-electrical conversion mechanism. The proposed energy harvester is composed of a generation cantilever and a bimetallic cantilever and uses the bimetallic effect of the bimetallic cantilever to achieve heat event-driven cycling energy harvesting. The temperature threshold of the energy harvester can be preset by adjusting the distance between the generation cantilever and the bimetallic cantilever. In the following sections, a technical description of the working principal, threshold design, and experimental results of the proposed energy harvester is given exhaustively.

## 2. Design and Modeling

### 2.1. Working Principle

The energy harvester consists of two miniature cantilevers that are electricity generation cantilever (GC) and bimetallic cantilever (BC). With lead zirconium titanate (PZT) piezoelectric film coated for generation, the highly thermo-conductive generation cantilever is magnetically coupled with the temperature sensitive bimetallic cantilever to realize a temperature threshold triggering effect, where the magnetized bimetallic cantilever is attracted to the end of the generation cantilever. The metal generation cantilever can be used to absorb or conduct heat from a heat source like fire or high temperature to cause spontaneous combustion. As shown in Figure 1a, the bimetallic cantilever is originally attracted into contact with the generation cantilever via an anchored graphite ring for flexible contact and good heat conductivity. The ferromagnetic bimetal structure of the bimetallic cantilever consists of two alloy layers with different thermal expansion coefficients. Graphite sheets are attached to the double surfaces of the bimetallic cantilever to enhance thermal transfer in endothermic process and heat radiation in exothermic process.

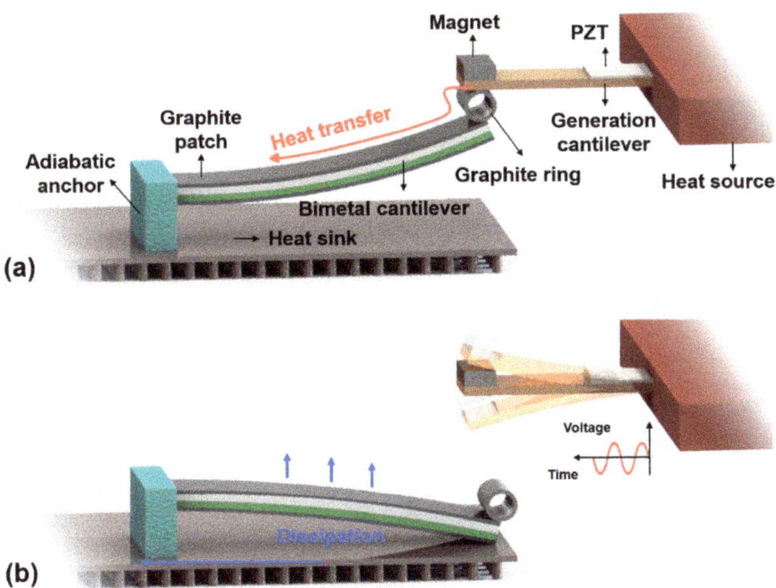



**Figure 1.** Schematic of the energy harvester for autonomous heat monitoring. (**a**) Endothermic state where the generation cantilever and the bimetallic cantilever are attracted; (**b**) exothermic state where the bimetallic cantilever and the generation cantilever are separated, and the freed generation cantilever is vibrating and generating electricity.

After the generation cantilever absorbs sufficient heat from the monitored object, the temperature of the bimetallic cantilever will rise quickly due to the high thermal conductivity of graphite. Then the bimetallic cantilever will bend downwards due to bimetallic effect, thereby causing deflection of the generation cantilever. The more the absorbed heat, the larger the deflection of the generation cantilever. When the deflection induced elastic force is greater than the magnetic force, the generation cantilever will get rid of the bimetallic cantilever and freely resonates to generate electricity with the piezoelectric film, as shown in Figure 1b. At this stage, the generation cantilever is in the state of end-free beam for free resonating power-generation. The critical temperature point of the heating induced separation between the generation cantilever and the bimetallic cantilever is defined as a temperature threshold. At the temperature threshold, the bimetallic cantilever is separated from the heat source and its bending causes contact with the heat sink metal plate placed below the bimetallic cantilever, and the heat in the bimetallic cantilever will be dissipated not only by thermal convection and radiation with environment but also through the contact with the heat sink. With the loss of heat, the bimetallic cantilever will gradually recover to its original position and be attracted to the generation cantilever. Then the heat dissipation induced bimetallic restoration of the bimetallic cantilever will cause the generation cantilever to vibrate and generate electricity again.

During the endothermic process induced electric harvesting, the generation cantilever is separated from the bimetallic cantilever, and the end-free cantilever vibrates to generate electricity. In contrast, during the exothermic process induced harvesting, the generation cantilever and the bimetallic cantilever are attracted together to form a double-clamped beam like vibrating structure. Therefore, the end-free cantilever features larger resonating amplitude and higher electricity-power generation compared to the double-clamped like structure, and the resonant frequency of the end-free cantilever is lower than that of the double-clamped like one. Thus, the power generation process caused by separation of the two cantilevers is called "strong harvest", and the generation caused by attraction between the two cantilevers is called "weak harvest". The generation cycle can be repeated to generate electric power and the electric generation itself signifies that the monitored heat event occurs, i.e., the heat induced temperature on the generation cantilever exceeds the temperature threshold that can be preset by specifically designing the structure.

*2.2. Thermal Simulation of the Bimetallic Cantilever*

According to the working principle of the proposed energy harvester, the response time of the energy harvester to temperature is quite dependent on the time required for the bimetallic cantilever to separate from the generation cantilever when heated, that is, the rate of heat conduction on the bimetallic cantilever. The response cycle of the system also depends heavily on the time required for the bimetallic cantilever to return to the original position after being separated from the heat source, that is, the heat dissipation rate of the bimetallic cantilever. Therefore, in order to improve the response rate of the entire system, this research mainly focuses on the effect of the temperature response rate of some attached materials that can be used for heat conduction/heat dissipation on the bimetallic cantilever.

There are three ways of heat transfer [1]: heat conduction, heat convection, and heat radiation. For this device, heat conduction plays a significant role in the process of the bimetallic cantilever absorbing heat to separate from the generation cantilever, because the better the heat conduction performance, the faster the bimetallic cantilever responds to the heat event. According to Fourier's Law expressed as Equation (1), it can be known that the performance of heat conduction is related to the thermal conductivity $\kappa$ of the thermally conductive material, the surface area $A$ crossed by the heat flux $Q$, the thickness $e$ of the heat transfer direction, and the temperature difference $\Delta T$. Thermal conductivity is a coefficient related to the properties of the material itself which means under the same operating conditions, the thermal conductivity mainly depends on the properties of the material itself.

$$Q = \kappa \times A \times \frac{\Delta T}{e} \tag{1}$$

According to the Newton's law of cooling expressed as Equation (2), the heat convection mainly depends on the temperature difference $\Delta T$ and the convective heat transfer coefficient $h_c$. The convective heat transfer coefficient mainly depends on the surface condition of the material and the physical properties of fluid. When the environmental conditions remain unchanged and the surface conditions of different thermal conductive materials are relatively consistent, $h_c$ can be regarded as a constant.

$$Q = h_c \times A \times \Delta T \tag{2}$$

Regarding the heat of thermal radiation, it can be analyzed by the Stefan–Boltzmann law shown in Equation (3):

$$Q = \varepsilon \times \sigma \times A \times \left( T^4 - T_a{}^4 \right) \tag{3}$$

where $\varepsilon$ is emissivity of the material, $\sigma$ is blackbody radiation constant which equals to $5.67 \times 10^{-8}$ W/(m²·K⁴), and $T_a$ is ambient temperature. From Equation (3) we can see that thermal radiation capacity not only depends on the emissivity but also heavily on the temperature of the object, since the exponent of temperature is the fourth power. The emissivity, similar to the convective heat transfer coefficient, is also a physical quantity related to the surface condition of the object. Therefore, it is necessary to find a kind of material with good thermal conductivity and excellent performance in the heat dissipation.

Due to the poor thermal conductivity of the bimetallic beam itself, as shown in Table 1, this study aimed at several common heat conductive materials on the market, such as graphite sheets, copper foils, and liquid alloy gallium-based alloys for simulation by using the Multiphysics software COMSOL. Table 1 shows the thermal conductivity and the emissivity of the aforementioned materials compared with the bimetal we used in the experiment.

**Table 1.** Thermal conductivity and emissivity of thermal conductive materials.

| Materials | Thermal Conductivity W/(m·K) | Emissivity |
| --- | --- | --- |
| Bimetal | 10.9 | 0.25 |
| Graphite sheet | 1200 | 0.85 |
| Copper foil | 400 | 0.3 |
| Gallium-based alloy | 120 | 0.5 |

The result of the simulation is shown in Figure 2. Figure 2a is a schematic diagram of the bimetallic cantilever after one end is fixed and the other end is deformed due to heat. The bimetallic cantilever has an initial curvature (8.33/m) at room temperature, so when one end is fixed, the other end will have an initial deflection. Figure 2b shows the transient heat conduction simulation when different thermal conductive materials are attached to the surface of the bimetallic cantilever. It can be seen from Figure 2b that the bimetallic cantilever with graphite sheets attached takes the shortest time to reach the highest steady-state temperature while the bimetallic cantilever without any thermal conductive materials attached has the longest heating time for the temperature to rise to its steady-state temperature, which indicates that attaching heat conductive materials can increase the thermal conductivity and steady-state temperature of the bimetallic cantilever. Figure 2c shows the simulation of heat dissipation of the bimetallic cantilever after it is separated from the heat source. It can be seen from Figure 2c that the bimetallic cantilever with graphite sheets still has the fastest heat dissipation rate, and the bimetallic cantilever without attaching any thermal conductive material has the second highest heat dissipation rate. The main reason for this phenomenon is that the attachment of the thermal conductive material increases the thermal resistance of the bimetallic cantilever when dissipating heat, which makes the heat dissipation capacity weaker than that of the no material attached situation. At the same time, the emissivity of the graphite sheet is much higher than the other thermal conductive materials, so the final heat dissipation capacity of the bimetallic cantilever with graphite sheet attached is the best.

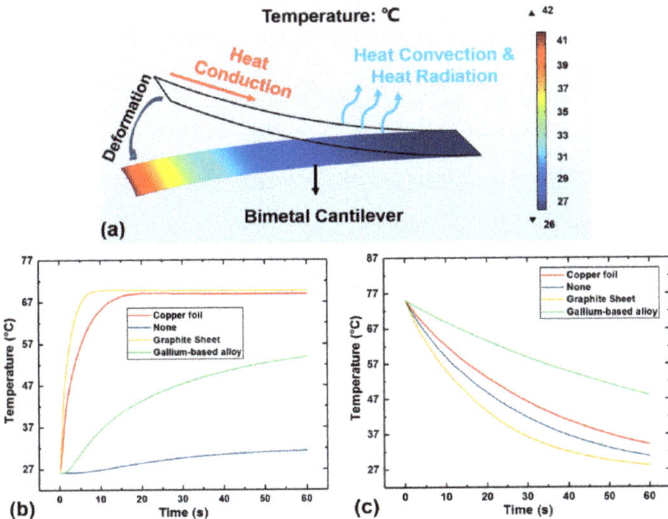


**Figure 2.** Heat conduction/dissipation simulation of the bimetallic cantilever. (**a**) Schematic diagram of the thermal deformation of the bimetallic cantilever; (**b**) the temperature curves when attaching different thermal conductive materials; and (**c**) the heat dissipation curves when attaching different thermal conductive materials.

*2.3. Temperature Threshold Simulation*

Since the low thermal expansion layer of the bimetallic cantilever is Invar alloy with nickel content of 36%, the bimetallic cantilever can be magnetized by the permanent magnet at the end of the generation cantilever. Therefore, the temperature threshold of the entire device can be set by the magnetic coupling between the two cantilevers, that is, the stronger the magnetic force, the higher the temperature threshold. The magnetic force between the generation cantilever and the bimetallic cantilever can be adjusted by the distance between the two stages. Table 2 shows the material properties and geometric parameters of the designed energy harvester.

**Table 2.** The material properties and geometric parameters of the energy harvester.

| Material Properties | Parameters |
| --- | --- |
| Bimetallic cantilever material | $Mn_{75}Ni_{15}Cu_{10}$ $Ni_{36}$ |
| Generation cantilever material | beryllium bronze |
| Magnet material | samarium cobalt |
| Maximum working temperature of the magnet | 350 °C |
| Bimetallic cantilever thickness | 0.1 mm |
| Bimetallic cantilever length | 47 mm |
| Bimetallic cantilever width | 10 mm |
| Generation cantilever thickness | 0.1 mm |
| Generation cantilever length | 18 mm |
| Generation cantilever width | 10 mm |
| Magnet length | 10 mm |
| Magnet width | 2 mm |
| Magnet height | 3 mm |
| PZT thickness | 0.06 mm |
| PZT length | 6 mm |
| PZT width | 10 mm |
| Distance between the two cantilevers | 2.5 mm |
| Graphite sheet thickness | 0.07 mm |

Figure 3a shows the magnetic force of the bimetallic cantilever magnetized by a permanent magnet changes with the distance between two cantilevers. According to the conclusion given in the literature [17], the relationship between magnetic force and distance can be expressed by the fourth power of the distance. Therefore, in this study, the curve obtained in Figure 3a is fitted with the reciprocal of the fourth-order polynomial of the distance $d$, and the coefficient of determination R-Square is 0.99997, indicating the Goodness of Fit meets the demand. Then, the relationship between the magnetic force and the distance between the two stages can be expressed as:

$$F_{mag} = \frac{1}{a_4 d^4 + a_3 d^3 + a_2 d^2 + a_1 d + a_0} \quad (a_4 = -0.09695, a_3 = -0.18457, a_2 = 2.11676, a_1 = -0.42567, a_0 = 1.09175) \quad (4)$$

Figure 3b shows the deflection of the generation cantilever with PZT attached based on Hooke's law. From Figure 3b, the equivalent stiffness $k_{eff}$ of the generation cantilever can be obtained and expressed as Equation (5):

$$F_{elastic} = k_{eff} \times z \quad (5)$$

where $z$ is the deflection of the generation cantilever, $k_{eff} = 122.0335$ N/m.

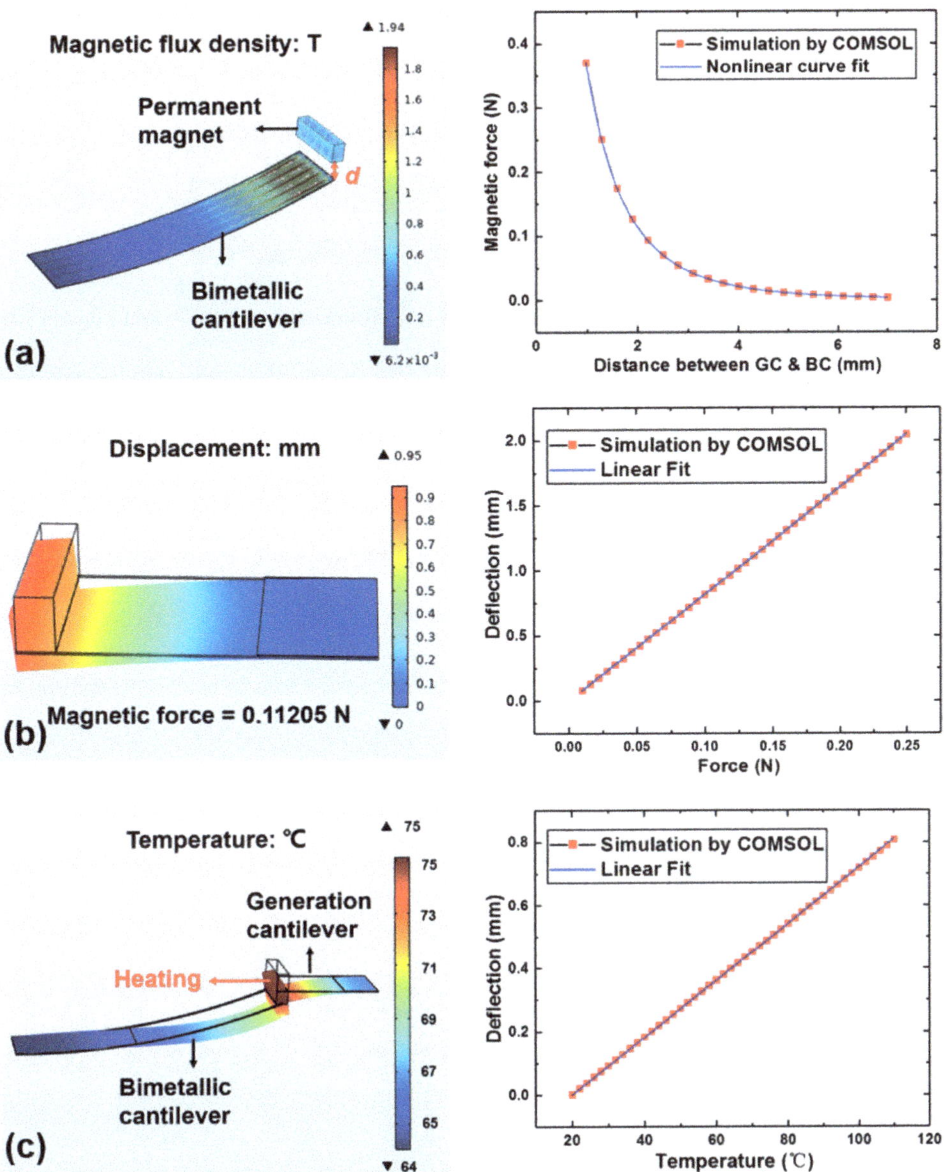



**Figure 3.** Simulation of the relationship between the temperature threshold and the distance between two stages. (**a**) The magnetic force on the magnetized bimetallic cantilever varies with distance; (**b**) deflection of the generation cantilever when the end of the cantilever is under force; (**c**) the deflection of the generation cantilever changes with the temperature, when the two cantilevers are attracted.

When the elastic force is greater than the magnetic force, the generation cantilever is separated from the bimetallic cantilever. Therefore, the situation when the elastic force is equal to the magnetic force can be regarded as the critical point. By combining Equations (4) and (5), the relationship between the deflection of the generation cantilever at critical point and the distance between two stages can be obtained. This can be described by Equation (6):

$$z = \frac{1}{k_{eff}} \times F_{mag} \tag{6}$$

Figure 3c shows the simulation results of the deflection of the generation cantilever change with temperature when the generation cantilever and the bimetallic cantilever are in attracted state. The area where the magnet is placed on the generation cantilever is the heated area. COMSOL simulation is conducted for the deflection of the bimetallic cantilever caused by different heat transfer methods shown in Figure 2a. If only the heat conduction is taken into account, the deflection would be 26.73 mm at 70 °C, and when heat convection and heat radiation are included into the simulation, the deflection is 23.40 mm, i.e., they both contribute about 12% to the total deflection. Herein, in order to improve the accuracy of simulation results, the convective heat transfer and radiation heat transfer with the environment are also considered in this model. Taking into account that when the model is working at the ambient temperature, the entire device should not generate any deformation, that is, when $T = T_a$, the deflection $z$ should be equal to zero. Meanwhile, it can be seen from the simulation results that the deflection $z$ has a linear relationship with the heated temperature $T$ which is consistent with the relationship between the deflection of the bimetallic cantilever and the heating temperature [18]. Therefore, the results obtained by simulation can be linearly fitted based on the above conclusions, and finally the relationship can be expressed as Equation (8), where $b = 0.00902$.

$$z = b \times (T - T_a) \tag{7}$$

Let Equation (7) be equal to the Equation (6), and the relationship between the temperature threshold and the distance between the two stages can be obtained and expressed as:

$$T = \frac{1}{b} \times \frac{1}{k_{eff}} \times F_{mag} + T_a \tag{8}$$

With the above relationship, the temperature threshold of the heat event autonomous monitoring energy harvester can be preset by adjusting the distance between the two stages. Figure 4 is the plot of the relationship between the temperature threshold and the distance between the two stages of the energy harvester according to Equation (8). It can be seen from Figure 4 that when the distance between the two stages is greater than 6 mm, the temperature threshold of the energy harvester is quite close to room temperature 20 °C, which means that there is almost no threshold between the two stages.

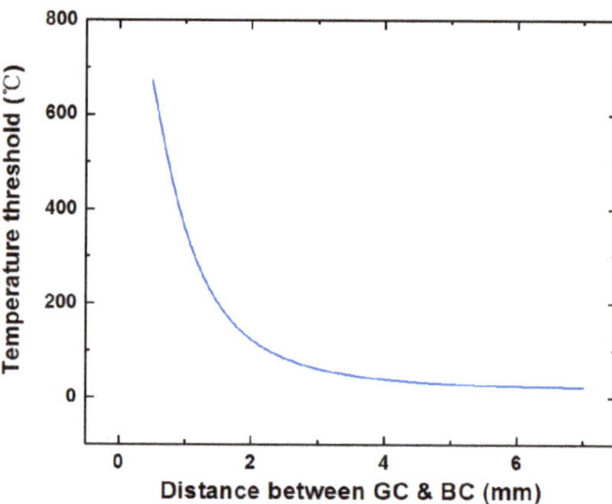


**Figure 4.** The relationship between the temperature threshold and the distance between the two stages of the energy harvester.

### 3. Experimental Results

Figure 5a is the test setup of the heat event induced energy harvester. For experimental convenience, the heat source is imitated by input direct current through an attached heating resistor (resistivity of 374 $\Omega/\text{m}$, resistance of 200 $\Omega$) on the generation cantilever. The current generated by DC source (Maisheng, MS1510D) can be adjusted by a PID temperature controller. Since it is essential to obtain the power generation data and the temperature data of the device at the same time, two kind of data acquisition units need to be used here. One data acquisition unit (NI USB-6003, 16 bit) with sampling rate as 1000 Hz is used to collect instantaneous analog signals from the generation cantilever, and the other (HRF USB-6009, 16 bit) with sampling rate as 0.5 Hz is used to collect device temperature data in real time. Both kinds of the data acquisition units can display the collected data through the display device. As shown in Figure 5b, the PZT sheet is bonded with the substrate through a kind of conductive silver paste (EPO TEK H20E). The end of the generation cantilever is equipped with a samarium cobalt magnet, whose working temperature can reach up to 350 °C which ensures the stability of the temperature threshold. The heating resistor is fixed near the end of the generation cantilever and an ultra-thin commercial K-type thermocouple with a wire diameter of 0.1 mm is put in touch with the generation cantilever in the same place to measure the heating temperature. The bimetallic cantilever consists of bimetallic layers of a high thermal expansion layer $Mn_{75}Ni_{15}Cu_{10}$ and a low expansion layer $Ni_{36}$. Thus, the bimetallic cantilever is ferromagnetic and can be magnetized by the magnet and attracted to the end of the generation cantilever. In order to enhance thermal conduction rate of the bimetallic cantilever, the graphite sheets are attached to the double surfaces of the bimetallic cantilever whose in-plane thermal conductivity can reach up to 1200 $W/(\text{m·k})$ which is about three times that of copper. Another ultra-thin commercial K-type thermocouple is placed at the end of the bimetallic cantilever to collect temperature data of the bimetallic cantilever. The graphite ring is placed between the generation cantilever and the bimetallic cantilever to increase the thermal contact area and has the function of adjusting the magnetic force between the two stages. In this way, the temperature threshold of the entire energy harvester can be adjusted by changing the size of the graphite ring. By both systematic analysis and experiment, the temperature threshold of the prototype device is set near 71 °C.

In our experiment, the bimetallic cantilever and the generation cantilever are always attracted together, that is, the generation cantilever never generates electricity, as long as the temperature of the heated generation cantilever is below 71 °C. However, when the temperature reaches or exceeds the threshold temperature of near 71 °C, the generation cantilever will get rid of the attraction and generate electricity through free vibration.

Figure 6a shows the temperature of the heating part and the temperature of the bimetallic cantilever change along with the separation and attraction process between the generation cantilever and the bimetallic cantilever. Initially, the two cantilevers are in attraction state, when the temperature of the bimetallic cantilever rises to no more than 42 °C, the two stages maintain the attracted state. At this time, the temperature of the heating part will drop from 75 °C to 71 °C due to heat flowing to the bimetallic cantilever. When the temperature at the end of the bimetallic cantilever exceeds 42 °C, the two cantilevers will be separated due to deflection of the bimetallic cantilever. In the separation state, the temperature of the bimetallic cantilever gradually decreases because of heat dissipation, and at the same time the temperature of the heating part of the generation cantilever can further rise. When the temperature at the end of the bimetallic cantilever drops to 32 °C and the temperature at the heating part rises to 75 °C, the bimetallic cantilever will restore deformation and return to its original position, and then the energy harvester will get into the attracted state again.

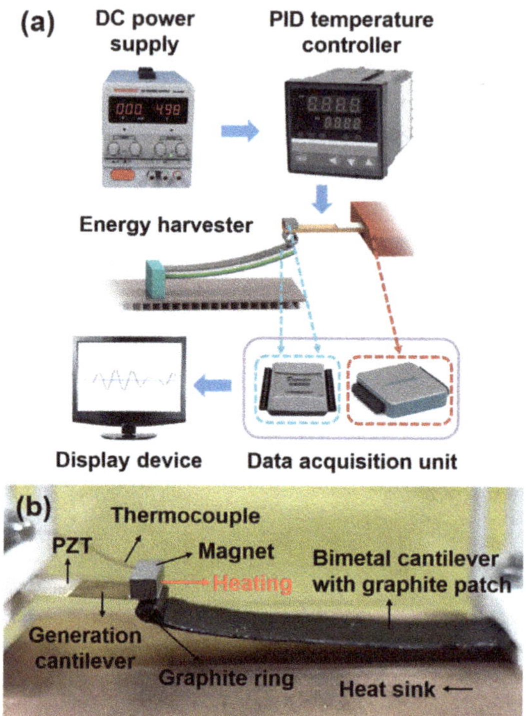



**Figure 5.** Experimental details of the heat event induced energy harvesting. (**a**) Schematic test setup of the energy harvester and (**b**) physical photo of the energy harvester prototype.

It should be noted that after the separation of the generation cantilever and the bimetallic cantilever, the heating part will be heated up to 75 °C; however, the energy harvester is first triggered at 71 °C, that is, the temperature of the two cantilevers when they are first separated is recognized as the threshold temperature which is 71 °C in this experiment.

Figure 6b shows the periodic power generation characteristics of the energy harvester for heating to 71 °C. A strong harvest process and a weak harvest process are observed in a cycling period of about 30 s. The strong harvest process occurs when the generation cantilever is pulling away from the bimetallic cantilever due to heat absorption, and this causes the generation cantilever to freely resonate and generate electric energy in the form of a one-end-free beam. The weak harvest process corresponds to the generation cantilever impacted and attracted again by the bimetallic cantilever. Due to heat dissipation deformation of the bimetallic cantilever is restored and returned to its original position, and this causes the generation cantilever to vibrate and generate electricity in the form of a buckled beam.

Figure 6c,d are the harvested voltage waveforms during the weak and strong process, respectively. The vibration frequency of the strong process is 34 Hz, which is smaller than 64 Hz during the weak process. The maximum instantaneous peak-to-peak voltage and the maximum instantaneous output power of strong harvest process are 1.12 V and 1.023 µW, respectively. The maximum instantaneous peak-to-peak voltage and the maximum instantaneous output power of weak harvest process are 0.34 V and 0.054 µW, respectively. The event of heating from ambient temperature (20 °C) to 71 °C causes the total instantaneous power generated during the whole cycle (including both the strong and weak harvest processes) reaches to 1.077 µW, and the electric energy generated in one cycle is 0.23 µJ when the temperature gradient is 51 °C.

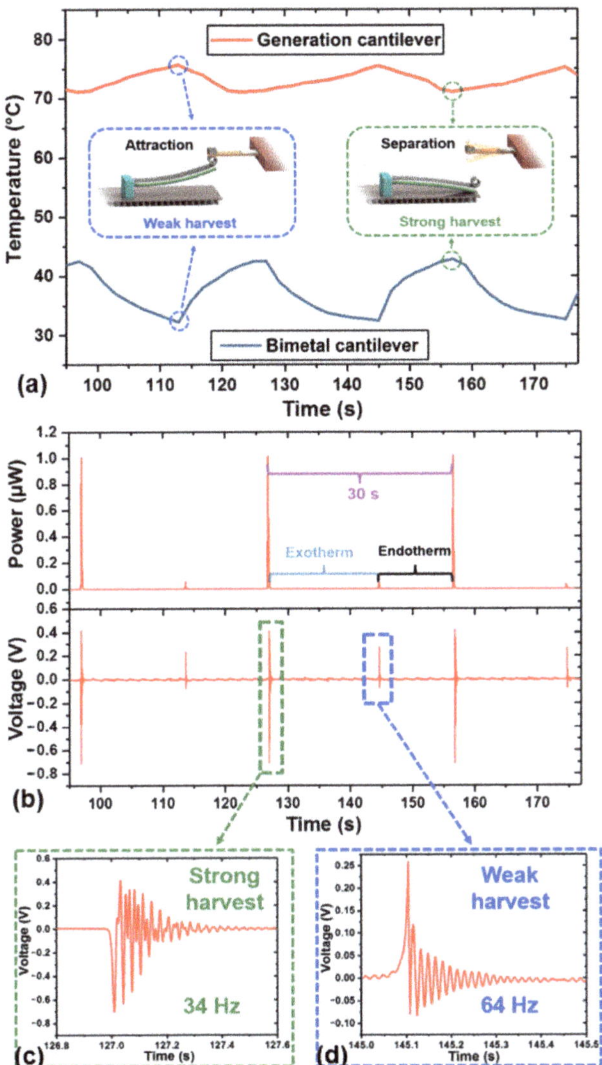



**Figure 6.** The temperature curves and power generation performance of the energy harvester when triggered at 71 °C. (**a**) The temperatures of the tip of the generation cantilever and the tip of the bimetal cantilever periodically change over time respectively; (**b**) instantaneous output power and output voltage periodically change over time respectively; (**c**) voltage waveform of the strong harvesting process; and (**d**) voltage waveform of the weak harvesting process.

Under similar temperature gradient, the proposed energy harvester may show lower power generation compared to the works in the literatures of [14,19]. However, our proposed device features the unique temperature-threshold triggered generation function that can be potentially used for autonomous event-driven autonomous sensing applications. In other words, the activity of electricity-power generating itself indicates the occurrence of the monitored object, i.e., the generated electricity also contains the sensing information of the monitored event. Unlike those conventional energy-harvesters, it is not necessary for this device to generate a sufficiently high electric-power to sustain the conventional sensors and their complicated interface-circuit systems. In an alternative method, the generated electric-energy in the proposed device can be only used to support alarming-signal sending out. This event-driven autonomous alarming principle means the device could possibly be used in monitoring WSN applications.

Figure 7 shows the tested peak-to-peak voltage and instantaneous output power of the 71 °C heating induced strong harvest process under different loading resistors. The maximum power of 0.72 µW was reached under the optimal load of 80 kΩ and the open-loop voltage was about 1.47 V. It should be noted that the test results of Figure 6 are all based on the optimal load resistance of 80 kΩ.

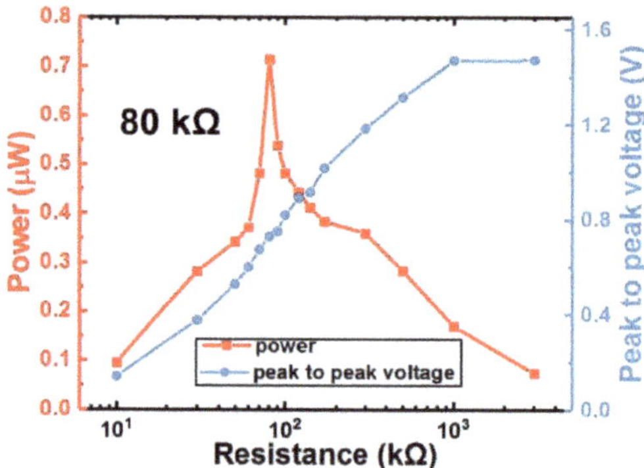


**Figure 7.** Test results of optimal loading resistance for the strong harvest.

Under different heating temperature of the event, Figure 8 shows the tested total generation period, the endothermic time, and the exothermic time which are both marked in Figure 6b after the device is triggered at 71 °C. Along with increasing heating temperature, the endothermic time required for the generation cantilever pulling away from the bimetallic cantilever gradually decreases. Correspondingly, the exothermic time required for the generation cantilever attracted by the bimetallic cantilever and the proportion of exothermic time to the entire generation period generally remain stable for the temperature range of 76 °C to 101 °C. Exceeding 101 °C however, the exothermic time greatly increases, which leads to an increase of the total time period of one cycle, possibly due to more heat needs to be dissipated when the working temperature exceeding 101 °C.

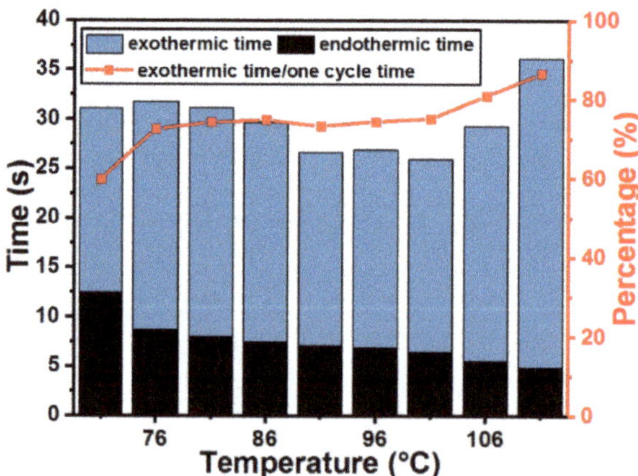


**Figure 8.** Tested endothermic and exothermic periods versus heating temperature.

## 4. Conclusions

A thermal energy harvester temperature threshold triggered generation for heat event autonomous monitoring is proposed and implemented. The proposed harvester can simultaneously output cyclic generation of electric energy and temperature-threshold triggered event information. In future work, the device would be further optimized to transmit the monitored heat event information by using the generated electric, thereby realizing self-powered WSN nodes for autonomous alarming.

**Author Contributions:** R.H. contributed to the whole work including design, fabrication, and testing of the device; N.W. contributed to the simulation analysis; Q.H. contributed to the research theoretical guidance; J.W. contributed to the testing guidance; and X.L. contributed to the research idea and gave guidance to the work. All authors have read and agreed to the published version of the manuscript.

**Funding:** The National Science Foundation of China (No. 61834007).

**Acknowledgments:** The authors appreciate financial support from the National Science Foundation of China Projects (61834007, 62074151, 61674160, 61527818) and the National Key Research and Development Program of the Ministry of Science and Technology of China (2016YFA0200803).

**Conflicts of Interest:** The authors declare no conflict of interest.

## References

1. Boughaleb, J.; Arnaud, A.; Cottinet, P.J.; Monfray, S.; Gelenne, P.; Kermel, P.; Quenard, S.; Boeuf, F.; Guyomar, D.; Skotnicki, T. Thermal modelling and optimization of a thermally matched energy harvester. *Smart Mater. Struct.* **2015**, *24*, 185–195. [CrossRef]
2. Olsson, R.H.; Bogoslovov, R.B.; Gordon, C. Event Driven Persistent Sensing: Overcoming the Energy and Lifetime Limitations in Unattended Wireless Sensors. In Proceedings of the 15th IEEE Sensors Conference, Orlando, FL, USA, 30 October–2 November 2016.
3. Kang, S.B.; Kim, J.H.; Jeong, M.H.; Sanger, A.; Kim, C.U.; Kim, C.M.; Choi, K.J. Stretchable and colorless freestanding microwire arrays for transparent solar cells with flexibility. *Light-Sci. Appl.* **2019**, *8*, 121. [CrossRef] [PubMed]
4. Liu, X.; Ma, J.; Wu, X.; Lin, L.; Wang, X. Polymeric nanofibers with ultrahigh piezoelectricity via self-orientation of nanocrystals. *ACS Nano* **2017**, *11*, 1901–1910. [CrossRef] [PubMed]
5. Wang, Y.; Shi, Y.; Mei, D.; Chen, Z. Wearable thermoelectric generator to harvest body heat for powering a miniaturized accelerometer. *Appl. Energy* **2018**, *215*, 690–698. [CrossRef]
6. Ravi, K.; Shashank, P. A review on low-grade thermal energy harvesting: Materials, methods and devices. *Materials* **2018**, *11*, 1433.
7. Yang, S.M.; Wang, S.H. Development of a thermoelectric energy generator with double cavity by standard cmos process. *IEEE Sens. J.* **2020**, *21*, 250–256. [CrossRef]
8. Hinterleitner, B.; Knapp, I.; Poneder, M.; Shi, Y.; Müller, H.; Eguchi, G.; Eisenmenger-Sittner, C.; Stöger-Pollach, M.; Kakefuda, Y.; Kawamoto, N.; et al. Thermoelectric performance of a metastable thin-film Heusler alloy. *Nature* **2019**, *576*, 85–90. [CrossRef] [PubMed]
9. Hao, F.; Qiu, P.; Tang, Y.; Bai, S.; Xing, T.; Chu, H.S.; Zhang, Q.; Lu, P.; Zhang, T.; Ren, D.; et al. High efficiency Bi2Te3-based materials and devices for thermoelectric power generation between 100 and 300 degrees C. *Energy Environ. Sci.* **2016**, *9*, 3120–3127. [CrossRef]
10. Hunter, S.R.; Lavrik, N.V.; Rajic, S.; Datskos, P.G. Review of pyroelectric thermal energy harvesting and new MEMs based resonant energy conversion techniques. In Proceedings of the Conference on Energy Harvesting and Storage: Materials, Devices, and Applications III, Baltimore, MD, USA, 23–24 April 2012; Spie-Int Soc Optical Engineering: Bellingham, WA, USA, 2012.
11. Leng, Q.; Chen, L.; Guo, H.; Liu, J.; Liu, G.; Hu, C.; Xi, Y. Harvesting heat energy from hot/cold water with a pyroelectric generator. *J. Mater. Chem. A* **2014**, *2*, 11940–11947. [CrossRef]
12. Kishore, R.A.; Priya, S. A review on design and performance of thermomagnetic devices. *Renew. Sustain. Energy Rev.* **2018**, *81*, 33–44. [CrossRef]
13. Chun, J.; Song, H.C.; Kang, M.G.; Kang, H.B.; Kishore, R.A.; Priya, S. Thermo-Magneto-Electric Generator Arrays for Active Heat Recovery System. *Sci. Rep.* **2017**, *7*, 41383. [CrossRef] [PubMed]
14. Deepak, K.; Varma, V.B.; Prasanna, G.; Ramanujan, R.V. Hybrid thermomagnetic oscillator for cooling and direct waste heat conversion to electricity. *Appl. Energy* **2019**, *233*, 312–320. [CrossRef]
15. Song, H.C.; Maurya, D.; Chun, J.; Zhou, Y.; Song, M.E.; Gray, D.; Yamoah, N.K.; Kumar, D.; McDannald, A.; Jain, M.; et al. Modulated Magneto-Thermal Response of $La_{0.85}Sr_{0.15}MnO_3$ and $(Ni_{0.6}Cu_{0.2}Zn_{0.2})Fe_2O_4$ Composites for Thermal Energy Harvesters. *Energy Harvest. Syst.* **2017**, *4*, 57–65. [CrossRef]

16. Waske, A.; Dzekan, D.; Sellschopp, K.; Berger, D.; Stork, A.; Nielsch, K.; Fähler, S. Energy harvesting near room temperature using a thermomagnetic generator with a pretzel-like magnetic flux topology. *Nat. Energy* **2019**, *4*, 68–74. [CrossRef]
17. Vokoun, D.; Beleggia, M.; Heller, L.; Sittner, P. Magnetostatic interactions and forces between cylindrical permanent magnets. *J. Magn. Magn. Mater.* **2009**, *321*, 3758–3763. [CrossRef]
18. Timoshenko, S. Analysis of bi-metal thermostats. *J. Opt. Soc. Am. Rev. Sci. Instrum.* **1925**, *11*, 233–255. [CrossRef]
19. Gao, Q.; Wang, W.; Lu, Y.; Cai, K.; Li, Y.; Wang, Z.; Wu, M.; Huang, C.; He, J. High Power Factor Ag/Ag$_2$Se Composite Films for Flexible Thermoelectric Generators. *ACS Appl. Mater. Interfaces* **2021**. [CrossRef] [PubMed]

*Article*

# Modeling of a Rope-Driven Piezoelectric Vibration Energy Harvester for Low-Frequency and Wideband Energy Harvesting

Jinhui Zhang, Maoyu Lin, Wei Zhou, Tao Luo and Lifeng Qin *

Department of Mechanical and Electrical Engineering, Xiamen University, Xiamen 361005, China; jinhuizhang@xmu.edu.cn (J.Z.); mylin@stu.xmu.edu.cn (M.L.); weizhou@xmu.edu.cn (W.Z.); luotao@xmu.edu.cn (T.L.)
* Correspondence: liq@xmu.edu.cn; Tel.: +86-134-0068-6106

**Abstract:** In this work, a mechanical model of a rope-driven piezoelectric vibration energy harvester (PVEH) for low-frequency and wideband energy harvesting was presented. The rope-driven PVEH consisting of one low-frequency driving beam (LFDB) and one high-frequency generating beam (HFGB) connected with a rope was modeled as two mass-spring-damper suspension systems and a massless spring, which can be used to predict the dynamic motion of the LFDB and HFGB. Using this model, the effects of multiple parameters including excitation acceleration, rope margin and rope stiffness in the performance of the PVEH have been investigated systematically by numerical simulation and experiments. The results show a reasonable agreement between the simulation and experimental study, which demonstrates the validity of the proposed model of rope-driven PVEH. It was also found that the performance of the PVEH can be adjusted conveniently by only changing rope margin or stiffness. The dynamic mechanical model of the rope-driven PVEH built in this paper can be used to the further device design or optimization.

**Keywords:** piezoelectric vibration energy harvester; low frequency; wideband; modeling

**Citation:** Zhang, J.; Lin, M.; Zhou, W.; Luo, T.; Qin, L. Modeling of a Rope-Driven Piezoelectric Vibration Energy Harvester for Low-Frequency and Wideband Energy Harvesting. *Micromachines* **2021**, *12*, 305. https://doi.org/10.3390/mi12030305

Academic Editor: Qiongfeng Shi

Received: 19 February 2021
Accepted: 13 March 2021
Published: 15 March 2021

**Publisher's Note:** MDPI stays neutral with regard to jurisdictional claims in published maps and institutional affiliations.

## 1. Introduction

In recent years, the great demand for micro-energy harvesting devices has been continuously increasing with the wide application of wearable device and wireless sensors due to the limited life-span and disposal pollution of the battery [1–3]. Mechanical vibration energy has been one of the major energy sources due to its ubiquity in the environment. The mechanisms for vibration energy harvesting can be mainly categorized into electromagnetic [4–6], electrostatic [7–9], piezoelectric [10–13], and triboelectric [14–17] ways. Since piezoelectric mechanism has a characteristic of simple structure, high mechanical-electrical conversion, and compatibility with CMOS (Complementary Metal Oxide Semiconductor), PVEH has been a hot research area in the past two decades for making self-powered sources to power small-scale systems [18–20]. For example, Kuang et al. proposed a PVEH combining magnets was designed to wear on the leg and could scavenge energy from knee-joint motions during human walking to provide sustainable energy supply for body sensors to realize energy-autonomous wireless sensing systems [21].

A traditional PVEH is typically composed of a piezoelectric cantilever beam and a proof mass attached to the free end of the cantilever. These kind of PVEHs usually have a high resonant frequency, especially for the micro PVEHs whose resonant frequency can reach thousands of hertz (shown in Table 1). Hence, it is difficult for the traditional cantilever beam type PVEH to be applied in many practical environment where the vibration frequency is quite low (<100 Hz, shown in Table 2).

**Table 1.** Typical micro piezoelectric vibration energy harvesters (PVEHs).

| References | Effective Volume (mm³) | Power (μW) | Acceleration (g) | Frequency (Hz) |
|---|---|---|---|---|
| Jeon et al. [22] | 0.027 | 1.01 | 10.8 | 13,900 |
| Renaud et al. [23] | 1.845 | 40 | 1.9 | 1800 |
| Shen et al. [24] | 0.6520 | 2.15 | 2.0 | 462.5 |
| Muralt et al. [25] | 0.48 | 1.4 | 2.0 | 870 |
| Elfrink et al. [26] | 15 | 69 | 0.2 | 599 |
| Park et al. [27] | 1.05 | 1.1 | 0.39 | 528 |
| Fang et al. [28] | 0.78 | 2.16 | 1.0 | 608 |
| Kanno.et al. [29] | 0.168 | 1.1 | 1.0 | 1036 |

**Table 2.** Common vibration sources [30,31].

| Vibration Sources | Acceleration (g) | Frequency (Hz) |
|---|---|---|
| Vanitation pipe | 0.02–0.15 | 60 |
| Lathe | 1.0 | 70 |
| Truck/Car engine | 0.052–0.198 | 37 |
| Human walking | 0.2–0.3 | 2–3 |
| Car instrument panel | 0.3 | 13 |
| Three-axis machine | 1.0 | 70 |
| Office building near the road | 0.02–0.15 | 60–100 |
| Tunnel train secondary vibration | 0.0026 | 15–25 |

Meanwhile, traditional PVEHs are usually single-mode resonant systems with relatively narrow bandwidth. Power output will be significantly reduced when the resonant frequency of a resonator and the ambient vibration frequency is mismatched. Therefore, it is a big challenge for PVEH to be used for in most practical applications, where the vibration frequency is time-variable. To address this issue, Leland and Wright proposed one technique to tune the resonant frequency of the harvester from 200 to 250 Hz by applying an axial compressive load since the resonant frequency of cantilever beam is stress dependent [32]. Similar research on resonant frequency shift by preload was also studied by Eichhorn et al., showing that one harvester could be altered from 380 to 292 Hz for a compressive preload, and another harvester was tuned from 440 to 460 Hz by applying a tensile preload [33]. However, the frequency tunable performance of these harvesters was achieved manually with the aids of preload mechanical structure. Hence, enabling the harvesters to automatically adjust its resonance frequency to match the driving frequency is still a challenge. Another strategy is to develop wideband PVEH whose working frequency can cover the frequency range of the vibration source as much as possible.

For achieving wideband energy harvesting, various methods have been proposed [34–36], which can be mainly categorized into multimodal technique and nonlinear techniques. As for multimodal technique, it usually uses an array of beams with different resonant frequencies or designing one beam to have multiple vibrational modals [37–39]. For example, Wang et al. [37] and Liu et al. [38] demonstrated that the array-beams PVEH could achieve wider bandwidth if the number of beams were added, where each beam had different frequency characteristics. Wu et al. [40] presented a PVEH based on one M-shaped beam comprising a main beam and two dimension-varied folded auxiliary beams interconnected through a proof mass at the end of the main cantilever. Such an energy harvester owns a three degree-of-freedom vibrating mode, and the resonant frequencies of its first three orders can be tuned close enough to obtain a utility wide bandwidth. Compared with tradition single-mode resonator system, the bandwidth of harvester can be greatly improved by multimodal mechanism, and wider or even unlimited bandwidth could be theoretically achieved if the number of beams continuously add for array-beams harvester. However, the disadvantage of multimodal technique is low volume efficiency because the fact that the bandwidth of single beam is not broadened.

On the contrary, the wideband energy harvesting using nonlinear techniques is increasing the bandwidth of single energy harvester based on nonlinear mechanism [35,41–44]. The most common nonlinear mechanism is the frequency up-conversion (FUC) technique, which can broaden the frequency bandwidth efficiently and offer a PVEH MEMS (Micro-Electro-Mechanical System) solution for low-frequency vibration energy harvesting. Generally, these frequency up-conversion (FUC) technologies can be classified into the contact and non-contact types. In terms of contact type, research has most commonly been conducted on the impact-driven PVEH that uses contact-mechanical force [11,43,45–48] to convert vibration energy to electrical energy. For example, Liu et al. [45] realized wideband energy harvesting and high output power via direct impact of a high-frequency piezoelectric $Pb(Zr_xTi_{1-x})O_3$ (PZT) beam using a low-frequency driving beam. Vijayan et al. [49] investigated a modified design in which a spring element is attached to one end of a beam, which can avoid damage to the device due to the direct mechanical impact between the hard beams. As for the harvesters using the non-contact FUC technique, non-contact forces such as an extra magnetic force [50–55] are used to convert vibration energy to electrical energy. For example, Tang et al. [52] proposed a bi-stable FUC PVEH that consists of a central sliding mass and a pair of piezoelectric generators. Permanent magnets are mounted on both the central mass and the ends of the generators. The sliding mass moves back and forth under the ambient excitation and intermittently repels generators to oscillate by the mutually repulsive magnetic force. Since it is non-contact frequency up-conversion, thereby avoiding mechanical collision and improving long-term working durability. Izadgoshsb et al. [56] presented a PVEH consisting of a double-pendulum and a PZT cantilever beam with magnets on their ends. By coupling the rotatable magnetic force interactions between the ends of the pendulum and PZT cantilever beam, the double-pendulum can impact the PZT cantilever beam for multiple times within one excitation period. Hence, the performance of the double-pendulum-based system in harvesting energy from low frequency human motions can be effectively improved compared to the conventional PVEH and single pendulum-based system. Although the nonlinear technique has been proven to be very efficient on broadening the bandwidth of harvester, the resulted bandwidth is still limited.

Hence, we proposed a low-frequency and wideband harvester based on a hybrid frequency up-conversion (FUC) nonlinear and multimodal mechanism in our previous work [57], aiming to combine the advantages of these two mechanisms, as demonstrated in Figure 1. The proposed wideband PVEH system (shown in Figure 1a) is composed of one high-frequency beam attached with piezoelectric material as the generating beam (HFGB) and multiple low-frequency beams with different frequencies as driving beams (LFDBs). Multiple LFDBs are connected mechanically with the HFGB using ropes. The main advantages of the rope-driven PVEH can be described as below:

(a)   Wider or even unlimited bandwidth could be achieved if the number of LFDBs are continuously increased. Unlike wideband PVEH based on an array piezoelectric beams in serial/parallel connection, the output performance of proposed PVEH will not deteriorate with the changing number of LFDBs, which has theoretically and experimentally been proved, and Figure 1b,c show a typical experimental result [57].

(b)   Similar to the impact-driven FUC wideband PVEH using a stopper, when an individual LFDB pulls the HFGB to oscillate it can achieve wideband energy harvesting, named as rope-driven FUC mechanism. Additionally, impact and rope-driven FUC mechanism can occur in the proposed PVEH by properly setting the length of rope, thus a much wider bandwidth could be achieved compared with the conventional impact-based FUC nonlinear wideband PVEH. Moreover, the working frequency of proposed PVEH can be tuned without re-fabricating or damaging the original structure by simply changing the rope length, which is ultra-convenient for practical applications. All these features of proposed PVEH has been experimentally verified, as shown in Figure 1d [58].

(c) Only HFGB is used for output in the proposed PVEH, which does not require a piezoelectric layer on LFDB, allowing great flexibility on the structure design of LFDB for various applications. For example, LFDB can adopted a curved shape shown in Figure 1e, which makes it easy to achieve low-frequency energy harvesting in a vibration environment, such as human motion, engine vibration, moving vehicles, and wave motion. Moreover, ultralow frequency, low intensity, and multidirectional vibration energy harvesting in a horizontal plane can be achieved if a liquid-based system is used as LFDB (see Figure 1f), which is difficult to be realized with traditional PVEHs [59].

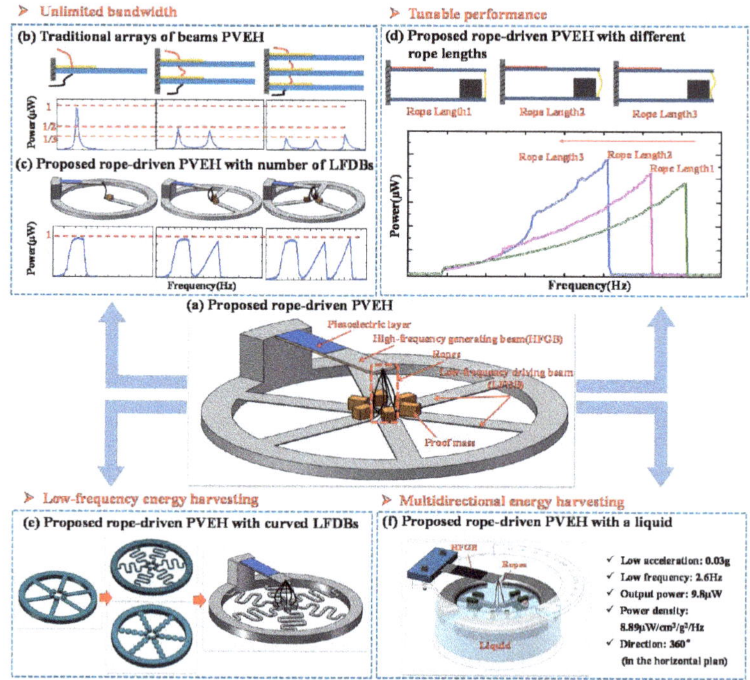


**Figure 1.** Schematic of the proposed novel low-frequency wideband piezoelectric energy harvester based on rope-driven mechanism for wideband, low-frequency, and multi-directional energy harvesting: (**a**) proposed rope-driven PVEH; (**b**) performance of traditional arrays of beams PVEH vs. number of beams; (**c**) performance of proposed rope-driven PVEH vs. number of LFDBs; (**d**) proposed rope-driven PVEH vs. rope lengths; (**e**) proposed rope-driven PVEH with curved LFDBs; (**f**) proposed rope-driven PVEH using liquid as energy capturing medium.

The mainly characteristics of the proposed PVEH have been demonstrated and discussed in detail in our prior work [57,58]. Whereas the dynamic process of LFBD driving HFGB is complicated, and multiple parameters such as acceleration, rope margin and rope stiffness, will greatly affect performance of harvester, which has not been systematically studied. Hence, in this paper, a rope-driven PVEH consisted of one LFDB and one HFGB is presented as a basic structure to systematically investigate the effects of the parameters in the performance of the PVEH by theoretical modeling and experimental verification. Here, a mechanical dynamic model is established based on mass-spring-damper systems, which can be used to predict the dynamic motion of the LFDB and HFGB effectively using lumped parameter modeling approach. The effects of multiple parameters, including acceleration, rope margin and rope stiffness, have been clarified systematically by numerical simulation and experiments using this model. The results show a reasonable agreement between the simulation and experimental study. This work will provide a basic mechanism understanding and give guidance for the design of rope-driven PVEH. The rest of the paper is organized as follows: Section 2 gives a mechanical model of the basic structure of rope-driven PVEH. Section 3 describes the experiment and simulation procedure. Section 4 mainly shows the parameter study on PVEH based on simulation and experiments, and Section 5 presents the conclusions drawn from the simulation and experimental results.

## 2. Modeling

The operation principle of the rope-driven PVEH with one LFDB and one HFGB is illustrated in Figure 2. The rope-driven PVEH is mounted on a base. Under a periodic excitation of the base, the LFDB pulls HFGB and transfers mechanical energy to HFGB by the rope for a short period in each vibration cycle when the LFDB is excited to exceed the margin $\Delta x$ (defined as length of rope $x_1$ minus initial distance of two beams $x_0$). This stage is named as Phase 1 (rope-driven vibration). This pull will give rise to a retardation of the LFDB's vibration amplitude but broaden the operating frequency bandwidth of LFDB. The reason is that HFGB arranged above the LFDB can be treated as a stopper just like the conventional impact-driven PVEH. As the rope pulls the HFGB, the effective stiffness of LFDB increases suddenly and results in a higher resonant frequency. Hence, the resonance of PVEH is extended over a wider interval of the frequency spectrum. After Phase 1, HFGB obtains vibrational energy from the LFDB and vibrates with exponentially attenuating amplitude at its higher resonant frequency (shown in Figure 2), and the cyclic deformation of piezoelectric material on HFGB will be transformed into electricity based on piezoelectric effect. LFDB continues to be excited by the base excitation until it pulls the HFGB in the next cycle. This phase is thought as Phase 2 (free vibration). The LFDB pulling the HFGB will not happen when the excitation is not large enough, and they will be forced to vibrate at the frequency of base excitation.

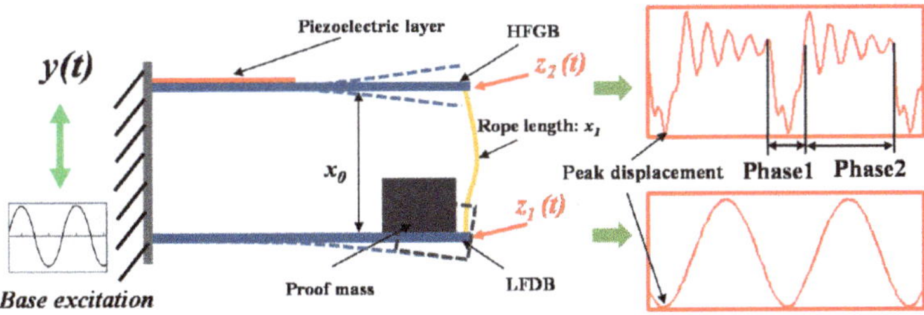


**Figure 2.** Schematic of the proposed PVEH.

A mechanical model of the rope-driven PVEH is established to describe the dynamic motion of PVEH base on the above mentioned working principle, as seen in Figure 3. Following the similar procedure [45], LFDB and HFGB can be modeled as two mass-spring-damper systems. LFDB, which is considered to be one suspension system, consists of a proof mass $m_1$ suspended by a spring $k_1$ and a damper $c_1$. HFGB is another suspension system, acting as a generator. Initially, HFGB is placed at a distance of $x_0$ above the LFDB, and has a damping coefficient $c_2$, stiffness $k_2$, and proof mass $m_2$. The base excitation $y(t)$ causes the proof mass $m_1$, $m_2$ to move relatively to the base movement ($y(t)$) as $z_1(t)$, $z_2(t)$, respectively. When the relative displacement of $z_1(t) - z_2(t)$ is bigger than rope margin $\Delta x$ (defined as length of rope $x_1$ minus initial distance of two beams $x_0$), the LFDB will pull HFGB by the rope. At this time, the rope can be treated as a massless spring with stiffness of $k_0$ considering the mass of rope is much less than the $m_1$, $m_2$ as shown in Figure 3a. Otherwise, when $z_1(t) - z_2(t)$ is smaller than $\Delta x$, the rope is ignored (see in Figure 3b).

Hence, corresponding to the mechanical model, the equations for dynamic motion of LFDB and HFGB's tip movement ($z_1(t)$, $z_2(t)$) can be built.

$$LFDB : \begin{cases} m_1\ddot{z}_1 + c_1\dot{z}_1 + k_1z_1 + k_0((z_1 - z_2) - \Delta x) = -m_1\ddot{y} & (z_1 - z_2 \geq \Delta x) \\ m_1\ddot{z}_1 + c_1\dot{z}_1 + k_1z_1 = -m_1\ddot{y} & (z_1 - z_2 < \Delta x) \end{cases} \quad (1)$$

$$HFGB : \begin{cases} m_2\ddot{z}_2 + c_2\dot{z}_2 + k_2z_2 - k_0((z_1 - z_2) - \Delta x) = -m_2\ddot{y} & (z_1 - z_2 \geq \Delta x) \\ m_2\ddot{z}_2 + c_2\dot{z}_2 + k_2z_2 = -m_2\ddot{y} & (z_1 - z_2 < \Delta x) \end{cases} . \quad (2)$$

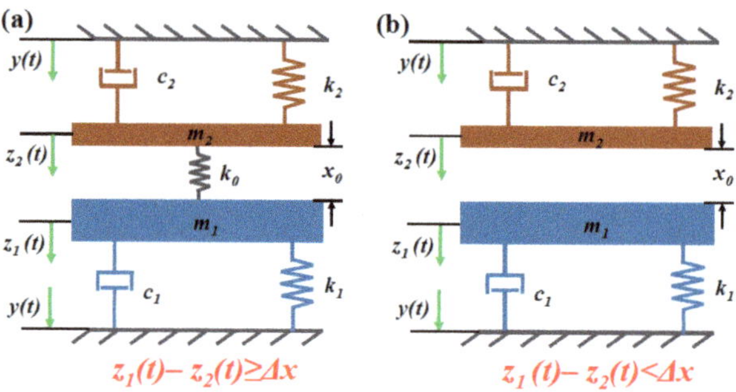


**Figure 3.** Mechanical model of the proposed PVEH as (a) $z_1 - z_2 \geq \Delta x$ and (b) $z_1 - z_2 < \Delta x$.

For simplifying the analysis, an assumption that beams and rope work on the linear range is adopted in the modeling. For harmonic base excitation $y(t) = Y\sin(\omega t)$, using $2\zeta_1\omega_1 = c_1/m_1$, $2\zeta_2\omega_2 = c_2/m_2$, $\omega_1{}^2 = k_1/m_1$ and $\omega_2{}^2 = k_2/m_2$, Equations (1) and (2) can be rewritten as follows:

$$LFDB : \ddot{z}_1 + 2\zeta_1\omega_1\dot{z}_1 + \omega_1{}^2z_1 = \begin{cases} -\frac{k_0}{m_1}((z_1 - z_2) - \Delta x) + a\sin\omega t & (z_1 - z_2 \geq \Delta x) \\ a\sin\omega t & (z_1 - z_2 < \Delta x) \end{cases} \quad (3)$$

$$HFGB : \ddot{z}_2 + 2\zeta_2\omega_2\dot{z}_2 + \omega_2{}^2z_2 = \begin{cases} +\frac{k_0}{m_2}((z_1 - z_2) - \Delta x) + a\sin\omega t & (z_1 - z_2 \geq \Delta x) \\ a\sin\omega t & (z_1 - z_2 < \Delta x) \end{cases} \quad (4)$$

where $\omega_1$ and $\zeta_1$ are the LFDB frequency and damping characteristics, $\omega_2$ and $\zeta_2$ are the HFGB frequency and damping characteristics, $a = Y\omega^2$ is the acceleration amplitude of the base excitation. $m_{1,2} = m_b/3 + m_p/3 + m$ [60], where $m_b$, $m_p$ and $m$ are the mass of substrate, piezoelectric layer, and proof mass. Thus, the displacements of the LFDB and HFGB versus the excitation frequency $\omega$ can be numerically derived based on Equations (3) and (4).

## 3. Experiment and Simulation Procedure

To determine the real-time displacement of LFDB and HFGB under different frequencies, a vibration monitoring system including a lock-in amplifier (STANFORD RESEARCH SYSTEM, Model SR830, Sunnyvale, CA, USA), vibration shaker (MB Dynamics, MODAL110, Cleveland, OH, USA), power amplifier (MB Dynamics, MB500VI, Cleveland, OH, USA), accelerometer (Baofei, JYD-2, Yangzhou, China), charge amplifier (Baofei, KD5002, Yangzhou, China), micro stages, two displacement sensors (KEYENCE, LK-G30/G10, Osaka, Japan) and a high-speed camera (PHANTOM, V611, Wayne, NJ, USA) were established, where the rope-driven PVEH system (see Figure 4) is fixed on a three-dimensional micro stage. In experiment, for the convenience of mechanical model vilification, LFDB and HFGB all use the pure beams without piezoelectric materials, and the detailed parameters are shown in Table 3. The damping ratios of LFDB and HFGB were measured from the exponentially decayed waveform using the relationship $\zeta = (1/2n\pi)\ln A_1/A_n$ [47], where $A_1$ and $A_n$ are the amplitudes of peaks. The rope stiffness $k_0$ is determined by the relationship $k_0 = EA/L$, where $E$ is young modulus measured by stress-strain relationship, $A$ and $L$ are cross-sectional area and length of rope determined by micrometer.

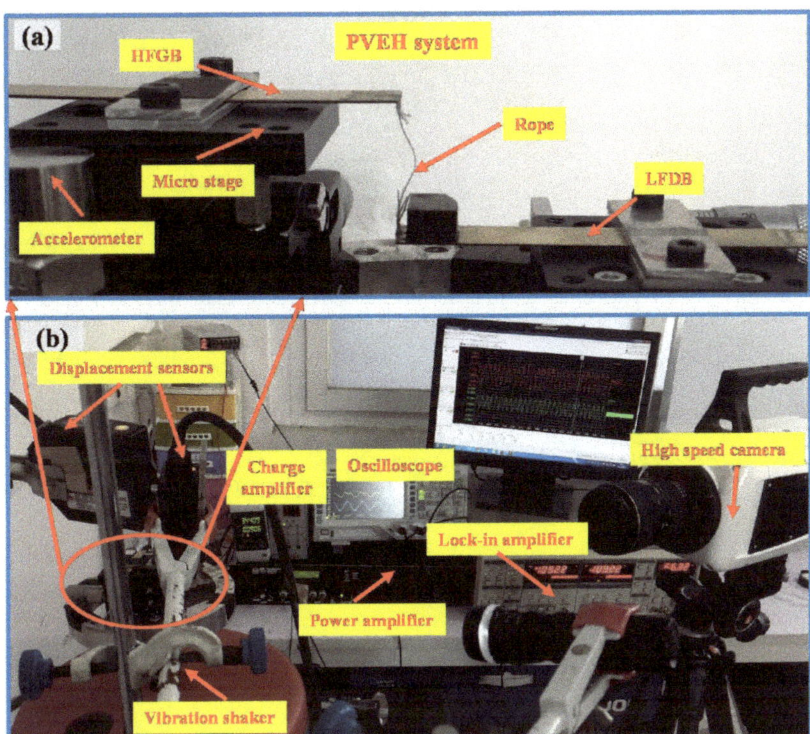


**Figure 4.** (**a**) Device fabrication and (**b**) detailed experimental setup of the PVEH system.

**Table 3.** Simulation and experiment parameters of LFDB, HFGB and the rope.

| Parameters | LFDB | HFGB | Rope |
|---|---|---|---|
| Length (mm) | 24.32 | 19.51 | 18.7 |
| Width (mm) | 6.00 | 6.00 | - |
| Thickness (mm) | 0.28 | 0.28 | - |
| Diameter (mm) | - | - | 0.1 |
| Proof mass (g) | 1.93 | 0.00 | - |
| Frequency (Hz) | 63.7 | 410.3 | - |
| Young modulus (Gpa) | 90 | 90 | 2.7 |
| Density(kg/m$^3$) | 8800 | 8800 | - |
| Damping ratio | $2.56 \times 10^{-3}$ | $5.79 \times 10^{-3}$ | - |

By adopting the parameters shown in Table 3, simulation results of LFDB and HFGB's displacements can be achieved using MATLAB/SIMULINK® model based on Equations (3) and (4). We firstly verify the feasibility of the mechanical model by comparing experimental results with simulation results. After that a parameter study was developed using the numerical simulation and experiments to understand the performance of the proposed PVEH. Specifically, we focus on the effect of rope, which is the easiest part for us to adjust once the beam structure is fabricated, in this paper.

## 4. Results and Discussion

Figures 5–8 show the simulation and experimental results of the proposed PVEH for acceleration *a* from 0.2 g to 0.6 g, rope margin $\Delta x$ from 0.5 to 1.5 mm and rope stiffness $k_0$ from 170 to 2500 N/m. The peak displacement of LFDB and HFGB is used as output to evaluate the characteristics of the rope-driven PVEH under different frequencies, which is defined as the maximum displacement during the rope-driven phase where LFDB pulls HFGB and reaches the lowest place in the downward direction (shown in Figure 2).

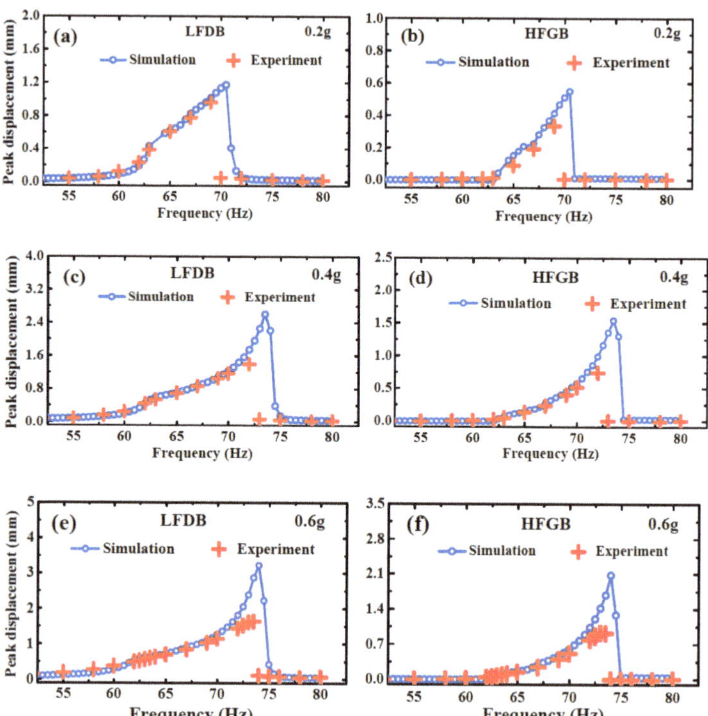

**Figure 5.** Experimental and simulation frequency responses of the peak displacement of LFDB with (**a**) *a* = 0.2 g, (**c**) *a* = 0.4 g and (**e**) *a* = 0.6 g; experimental and simulation frequency responses of the peak displacement of HFGB with (**b**) *a* = 0.2 g, (**d**) *a* = 0.4 g and (**f**) *a* = 0.6 g.

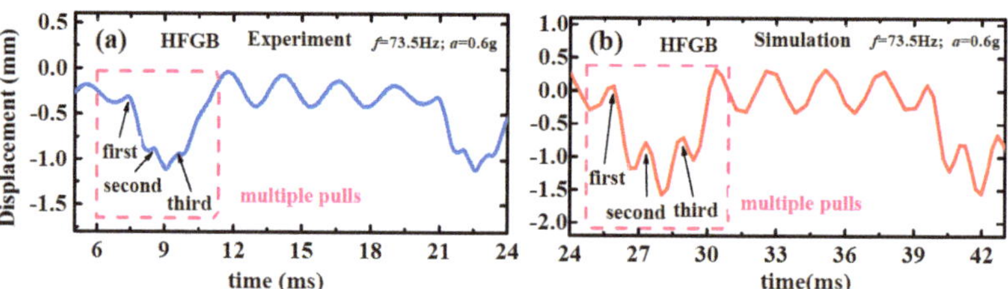

**Figure 6.** Typical real-time experimental (**a**) and simulation (**b**) displacement of HFGB.

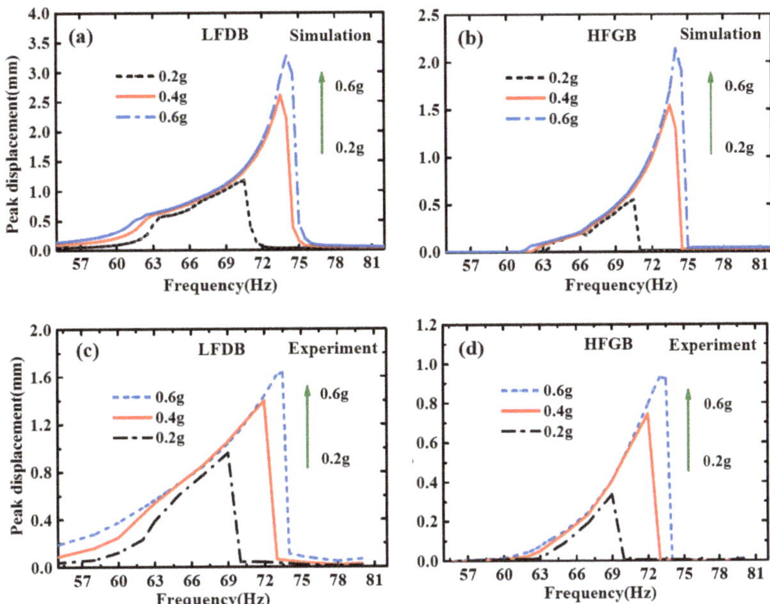

**Figure 7.** Frequency responses of the peak displacement of LFDB and HFGB for various accelerations under the condition of $\Delta x = 0.5$ mm and $k_0 = 1150$ N/m: (**a**) simulation frequency responses of LFDB; (**b**) simulation frequency responses of HFGB; (**c**) experimental frequency responses of LFDB; (**d**) experimental frequency responses of HFGB.

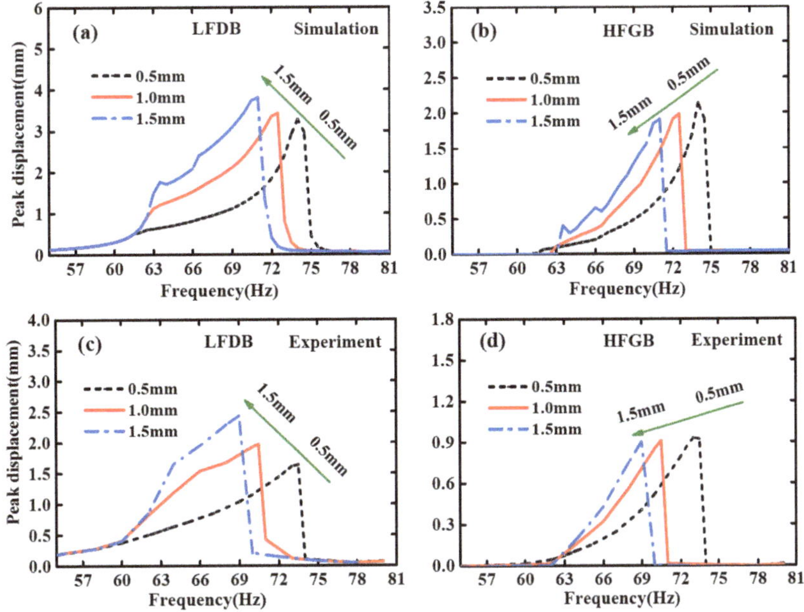

**Figure 8.** Frequency responses of the peak displacement of LFDB and HFGB for various rope-margins under the condition of $a = 0.6$ g and $k_0 = 1150$ N/m: (**a**) simulation frequency responses of LFDB; (**b**) simulation frequency responses of HFGB; (**c**) experimental frequency responses of LFDB; (**d**) experimental frequency responses of HFGB.

As seen in these figures, the simulation results basically agree well with the experimental results, demonstrating the validity of our rope-driven PVEH modeling though there are slight mismatches in amplitude over special frequency range. Figure 5 shows typical simulation and experimental frequency responses curves of LFDB and HFGB with $\Delta x = 0.5$ mm, $k_0 = 1150$ N/m, and $a = 0.2$ g, 0.4 g, 0.6 g. It can be seen that the simulation curve coincide the experimental curve in most frequency ranges, and the minimum error of displacement can reach 1.3%. However, the displacement discrepancy of simulation and experiment is more obvious when excitation frequencies get close to the end-frequency point (corresponding to the frequency point where the output suddenly drops to zero). Interestingly, real-time experimental and simulation displacement curves of HFGB (shown in Figure 6) demonstrated that shape of dynamic motion curves are still kept similar and has same pulling times. The mismatches in amplitude over the end-frequency range are probably due to two reasons. Firstly, the closer to the end-frequency point, the more strongly the whole PVEH system vibrates, which makes the beams and rope work on the nonlinear range possibly. However, the mechanical model we built gives no consideration to the nonlinear terms of beam movements. Furthermore, it can be seen that the higher acceleration is, the mismatch over the end-frequency range is more obvious by comparing Figure 5a,b with Figure 5c–f. Secondly, the rope is equivalent to a spring without damper in the established model, which may be oversimplified, considering the complicate movement of HFGB and LFDB in the multiple-pulling stage (shown in Figure 6). Hence, the non-linear terms of beams and the good equivalent of rope under strong excitation should be addressed for the model improvement in the future.

In general, the model established in this paper can be used to predict the characteristics of the rope-driven PVEH effectively. For example, in Figure 7, both simulation and experimental results show that wider bandwidth and higher output displacement of LFDB and HFGB can be observed when the excitation increases from 0.2 g to 0.6 g. Using the numerical simulation system in this paper, the characteristic of PVEH can be predicted qualitatively before fabricating the device for guiding the design of the PVEH.

According to the established mechanical model, we know that the performance of the PVEH is affected by multiple parameters, which mainly includes geometric parameters of LFDB and HFGB, margin and stiffness of rope, and excitation acceleration level. As the PVEHs applied in a specific application, once the PVEHs are fabricated, the characteristics of PVEHs will be fixed. Usually, owing to the fabrication error, there would be a different performance between the designed PVEHs and the fabricated PVEHs. If to change the performance of PVEHs, it will be difficult and inconvenient by changing the parameters of LFDB or HFGB, especially for the MEMS PVEHs, unless the devices are re-fabricated, while by adding the rope to PVEHs, only controlling the rope margin, the desired performance can be realized effectively. Hence, we mainly focus on the parameter study of rope margin and stiffness by using numerical simulation and experimental investigation.

Figure 8 shows the frequency responses of the peak displacement of LFDB and HFGB for different rope-margins. It can be seen from the simulation and experimental results that the performance of PVEH can be tuned by controlling the rope length. Taking the frequency responses curves in Figure 8b of HFGB as an example, as the rope margin changing from 1.5 mm to 0.5 mm, the bandwidth can be broadened from 7.5 Hz (63.5–71 Hz) to 12.5 Hz (62–74.5 Hz), although there is a slight drop in the output near the end-frequency point. Thus, changing rope could become a more convenient choice for adjusting the performance of the PVEHs applied in the real applications, and can also increase the tolerance of fabrication.

Figure 9 depicts frequency responses of the peak displacement of LFDB and HFGB for various rope stiffness $k_0$ changing from 1150 to 2500 N/m. Both simulation and experiments suggest that bandwidth and output displacement of LFDB and HFGB (shown in Figure 9a,b,e,f) are weakened but not much compared to the effect of rope margin when increasing the rope stiffness such as from 1150 to 2500 N/m. To further understanding the effect of rope margin, small rope stiffness less than 1150 N/m was investigated by simulation, which is shown in Figure 9c,d. In contrast to the conclusions as rope stiffness $k_0$

changing from 1150 to 2500 N/m, the bandwidth of LFDB increases when the rope stiffness is at a range of 170–1150 N/m, whereas its displacement decreases (see in Figure 9c). As for HFGB, the bandwidth increases, whereas the displacement hardly decreases (see in Figure 9d). It means that the bandwidth and output of rope-driven PVEH can also be adjusted by rope stiffness, and could be improved further if the stiffness of rope increases in a certain range. Overall, for practical applications, adjusting rope margin is a better and more effective way to change the rope-driven PVEH's performance to achieve tunable energy harvesting.

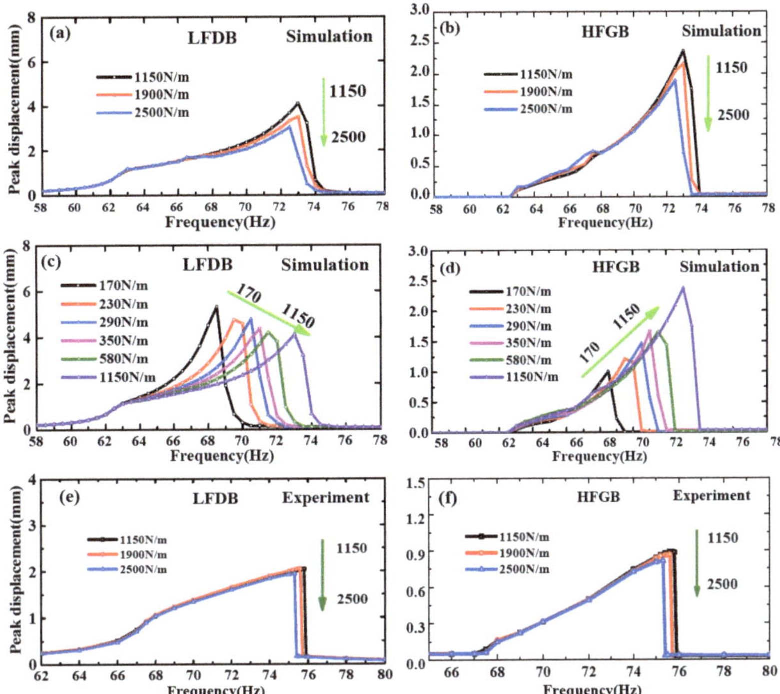

**Figure 9.** Frequency responses of the peak displacement of LFDB and HFGB for various rope stiffness under the condition of $a = 0.6$ g and $\Delta x = 1.0$ mm: (**a**) simulation frequency responses of LFDB under $k_0 = 1150$–2500 N/m; (**b**) simulation frequency responses of HFGB under $k_0 = 1150$–2500 N/m; (**c**) simulation frequency responses of LFDB under $k_0 = 170$–1150 N/m; (**d**) simulation frequency responses of HFGB under $k_0 = 170$–1150 N/m; (**e**) experimental frequency responses of LFDB under $k_0 = 1150$–2500 N/m; (**f**) experimental frequency responses of HFGB under $k_0 = 1150$–2500 N/m.

## 5. Conclusions

In conclusion, a mechanical mass-spring-damper model of the rope-driven PVEH with one LFDB connecting one HFGB by a rope was built using lumped parameter modeling approach, and the dynamic equations for predicting the dynamic movement of LFDB and HFGB were established, respectively. Based on the equations using MATLAB/SIMULINK® model, the effects of multiple parameters including excitation acceleration, rope margin and stiffness were investigated by numerical simulation and experiments. Overall, the results show that the established model in this study can be used to predict the behaviors of our rope-driven PVEH though there are some mismatches near the end-frequency point, which can provide guidance for the design of the rope-driven PVEH before fabricating it. The parameter study suggests that rope-driven PVEH's performance can be optimized by multiple parameters. Adjusting rope is the most convenient choice for optimizing the performance of the PVEH in variable environments without refabricating the device, this tunable performance of the proposed rope-driven PVEH system makes it promising for vibration energy harvesting in wideband environments with low frequency.

**Author Contributions:** Conceptualization, J.Z. and L.Q.; methodology, J.Z.; software, J.Z.; validation, J.Z. and M.L.; formal analysis, J.Z. and L.Q.; investigation, J.Z.; data curation, J.Z. and M.L.; writing—original draft preparation, J.Z.; writing—review and editing, L.Q., W.Z. and T.L.; visualization, J.Z.; funding acquisition, L.Q. and J.Z. All authors have read and agreed to the published version of the manuscript.

**Funding:** This work was supported by the National Natural Science Foundation of China (Grant No. 51775465, 52005423), Guangdong Basic and Applied Basic Research Foundation (Grant No. 2020A1515011486) and China Postdoctoral Science Foundation (Grant No. 2020M671946).

**Conflicts of Interest:** The authors declare no conflict of interest.

# References


1. Guan, M.J.; Liao, W.H. On the efficiencies of piezoelectric energy harvesting circuits towards storage device voltages. *Smart Mater. Struct.* **2007**, *16*, 498–505. [CrossRef]
2. Jiang, L.; Yang, Y.; Chen, R.; Lu, G.; Li, R.; Xing, J.; Shung, K.K.; Humayun, M.S.; Zhu, J.; Chen, Y.; et al. Ultrasound-Induced Wireless Energy Harvesting for Potential Retinal Electrical Stimulation Application. *Adv. Funct. Mater.* **2019**, *29*, 1–13. [CrossRef]
3. Yang, Y.; Hu, H.; Chen, Z.; Wang, Z.; Jiang, L.; Lu, G.; Li, X.; Chen, R.; Jin, J.; Kang, H.; et al. Stretchable Nanolayered Thermoelectric Energy Harvester on Complex and Dynamic Surfaces. *Nano Lett.* **2020**, *20*, 4445–4453. [CrossRef] [PubMed]
4. Sardini, E.; Serpelloni, M. An efficient electromagnetic power harvesting device for low-frequency applications. *Sens. Actuators A Phys.* **2011**, *172*, 475–482. [CrossRef]
5. Halim, M.A.; Park, J.Y. Modeling and experiment of a handy motion driven, frequency up-converting electromagnetic energy harvester using transverse impact by spherical ball. *Sens. Actuators A Phys.* **2015**, *229*, 50–58. [CrossRef]
6. Foisal, A.R.M.; Hong, C.; Chung, G.-S. Multi-frequency electromagnetic energy harvester using a magnetic spring cantilever. *Sens. Actuators A Phys.* **2012**, *182*, 106–113. [CrossRef]
7. Zhang, Y.; Wang, T.; Luo, A.; Hu, Y.; Li, X.; Wang, F. Micro electrostatic energy harvester with both broad bandwidth and high normalized power density. *Appl. Energy* **2018**, *212*, 362–371. [CrossRef]
8. Boisseau, S.; Despesse, G.; Ricart, T.; Defay, E.; Sylvestre, A. Cantilever-based electret energy harvesters. *Smart Mater. Struct.* **2011**, *20*, 105013–105023. [CrossRef]
9. Tao, K.; Lye, S.W.; Miao, J.; Tang, L.; Hu, X. Out-of-plane electret-based MEMS energy harvester with the combined nonlinear effect from electrostatic force and a mechanical elastic stopper. *J. Micromech. Microeng.* **2015**, *25*, 104014–104024. [CrossRef]
10. Jung, I.; Shin, Y.H.; Kim, S.; Choi, J.Y.; Kang, C.Y. Flexible piezoelectric polymer-based energy harvesting system for roadway applications. *Appl. Energy* **2017**, *197*, 222–229. [CrossRef]
11. Liu, H.; Lee, C.; Kobayashi, T.; Tay, C.J.; Quan, C. Piezoelectric MEMS-based wideband energy harvesting systems using a frequency-up-conversion cantilever stopper. *Sens. Actuators A Phys.* **2012**, *186*, 242–248. [CrossRef]
12. Liu, S.; Cheng, Q.; Zhao, D.; Feng, L. Theoretical modeling and analysis of two-degree-of-freedom piezoelectric energy harvester with stopper. *Sens. Actuators A Phys.* **2016**, *245*, 97–105. [CrossRef]
13. Zeng, Y.; Jiang, L.; Sun, Y.; Yang, Y.; Quan, Y.; Wei, S.; Lu, G.; Li, R.; Rong, J.; Chen, Y.; et al. 3D-printing piezoelectric composite with honeycomb structure for ultrasonic devices. *Micromachines* **2020**, *11*, 713. [CrossRef] [PubMed]
14. Ma, Z.; Ai, J.; Shi, Y.; Wang, K.; Su, B. A Superhydrophobic Droplet-Based Magnetoelectric Hybrid System to Generate Electricity and Collect Water Simultaneously. *Adv. Mater.* **2020**, 1–9. [CrossRef]
15. Cheng, T.; Gao, Q.; Wang, Z.L. The Current Development and Future Outlook of Triboelectric Nanogenerators: A Survey of Literature. *Adv. Mater. Technol.* **2019**, 1–7. [CrossRef]
16. Wang, Z.L. Triboelectric nanogenerators as new energy technology and self-powered sensors - Principles, problems and perspectives. *Faraday Discuss.* **2014**, *176*, 447–458. [CrossRef] [PubMed]
17. Wang, Y.; Yang, Y.; Wang, Z.L. Triboelectric nanogenerators as flexible power sources. *NPJ Flex. Electron.* **2017**, *1*, 1–10. [CrossRef]
18. Saadon, S.; Sidek, O. A review of vibration-based MEMS piezoelectric energy harvesters. *Energy Convers. Manag.* **2011**, *52*, 500–504. [CrossRef]
19. Wang, L.; Wang, Z.L. Advances in piezotronic transistors and piezotronics. *Nano Today* **2021**, *37*, 101108. [CrossRef]
20. Karan, S.K.; Maiti, S.; Lee, J.H.; Mishra, Y.K.; Khatua, B.B.; Kim, J.K. Recent Advances in Self-Powered Tribo-/Piezoelectric Energy Harvesters: All-In-One Package for Future Smart Technologies. *Adv. Funct. Mater.* **2020**, *30*, 1–52. [CrossRef]
21. Kuang, Y.; Ruan, T.; Chew, Z.J.; Zhu, M. Energy harvesting during human walking to power a wireless sensor node. *Sens. Actuators A Phys.* **2017**, *254*, 69–77. [CrossRef]
22. Jeon, Y.B.; Sood, R.; Jeong, J.H.; Kim, S.G. MEMS power generator with transverse mode thin film PZT. *Sens. Actuators A Phys.* **2005**, *122*, 16–22. [CrossRef]

23. Renaud, M.; Sterken, T.; Schmitz, A.; Fiorini, P.; Van Hoof, C.; Puers, R. Piezoelectric harvesters and MEMS technology: Fabrication, modeling and measurements. In Proceedings of the Solid-State Sensors, Actuators & Microsystems Conference, Transducers International, Lyon, France, 10–14 June 2007; pp. 891–894.

24. Shen, D.; Park, J.; Ajitsaria, J.; Choe, S.; Iii, H.C.W.; Kim, D. The design, fabrication and evaluation of a MEMS PZT cantilever with an integrated Si proof mass for vibration energy harvesting. *J. Microelectromech. Syst.* **2008**, *18*, 055017. [CrossRef]

25. Muralt, P.; Marzencki, M.; Belgacem, B.; Calame, F.; Basrour, S. Vibration Energy Harvesting with PZT Micro Device. *Procedia Chem.* **2009**, *1*, 1191–1194. [CrossRef]

26. Elfrink, R.; Renaud, M.; Kamel, T.M.; De Nooijer, C.; Jambunathan, M.; Goedbloed, M.; Hohlfeld, D.; Matova, S.; Pop, V.; Caballero, L.; et al. Vacuum-packaged piezoelectric vibration energy harvesters: Damping contributions. *J. Micromech. Microeng.* **2010**, *20*, 104001. [CrossRef]

27. Park, J.C.; Member, S.; Park, J.Y.; Lee, Y. Modeling and Characterization of Piezoelectric d33 Mode MEMS Energy Harvester. *J. Microelectromech. Syst.* **2010**, *19*, 1215–1222. [CrossRef]

28. Fang, H.B.; Liu, J.Q.; Xu, Z.Y.; Dong, L.; Wang, L.; Chen, D.; Cai, B.C.; Liu, Y. Fabrication and performance of MEMS-based piezoelectric power generator for vibration energy harvesting. *Microelectron. J.* **2006**, *37*, 1280–1284. [CrossRef]

29. Kanno, I.; Ichida, T.; Adachi, K.; Kotera, H.; Shibata, K. Power-generation performance of lead-free (K,Na)NbO$_3$ piezoelectric thin-film energy harvesters. *Sens. Actuators A Phys.* **2012**, *179*, 132–136. [CrossRef]

30. Roundy, S.; Wright, P.K.; Rabaey, J. A study of low level vibrations as a power source for wireless sensor nodes. *Comput. Commun.* **2003**, *26*, 1131–1144. [CrossRef]

31. Li, H.; Tian, C.; Deng, Z.D.; Li, H.; Tian, C.; Deng, Z.D. Energy harvesting from low frequency applications using piezoelectric materials. *Appl. Phys. Rev.* **2014**, *1*, 041301. [CrossRef]

32. Leland, E.S.; Wright, P.K. Resonance tuning of piezoelectric vibration energy scavenging generators using compressive axial preload. *Smart Mater. Struct.* **2006**, *15*, 1413–1420. [CrossRef]

33. Eichhorn, C.; Goldschmidtboeing, F.; Woias, P. Bidirectional frequency tuning of a piezoelectric energy converter based on a cantilever beam. *J. Micromech. Microeng.* **2009**, *19*. [CrossRef]

34. Liu, H.; Fu, H.; Sun, L.; Lee, C.; Yeatman, E.M. Hybrid energy harvesting technology: From materials, structural design, system integration to applications. *Renew. Sustain. Energy Rev.* **2020**, *137*, 1–25. [CrossRef]

35. Tran, N.; Ghayesh, M.H.; Arjomandi, M. Ambient vibration energy harvesters: A review on nonlinear techniques for performance enhancement. *Int. J. Eng. Sci.* **2018**, *127*, 162–185. [CrossRef]

36. Daqaq, M.F.; Masana, R.; Erturk, A.; Dane Quinn, D. On the Role of Nonlinearities in Vibratory Energy Harvesting: A Critical Review and Discussion. *Appl. Mech. Rev.* **2014**, *66*, 040801. [CrossRef]

37. Xue, H.; Hu, Y.; Wang, Q.M. Broadband piezoelectric energy harvesting devices using multiple bimorphs with different operating frequencies. *IEEE Trans. Ultrason. Ferroelectr. Freq. Control* **2008**, *55*, 2104–2108. [CrossRef]

38. Liu, J.Q.; Fang, H.B.; Xu, Z.Y.; Mao, X.H.; Shen, X.C.; Chen, D.; Liao, H.; Cai, B.C. A MEMS-based piezoelectric power generator array for vibration energy harvesting. *Microelectron. J.* **2008**, *39*, 802–806. [CrossRef]

39. Li, X.; Upadrashta, D.; Yu, K.; Yang, Y. Analytical modeling and validation of multi-mode piezoelectric energy harvester. *Mech. Syst. Signal Process.* **2019**, *124*, 613–631. [CrossRef]

40. Wu, M.; Ou, Y.; Mao, H.; Li, Z.; Liu, R.; Ming, A.; Ou, W. Multi-resonant wideband energy harvester based on a folded asymmetric M-shaped cantilever. *AIP Adv.* **2015**, *104*, 077149. [CrossRef]

41. Zhou, S.; Zuo, L. Nonlinear dynamic analysis of asymmetric tristable energy harvesters for enhanced energy harvesting. *Commun. Nonlinear Sci. Numer. Simulat.* **2018**, *61*, 271–284. [CrossRef]

42. Tang, L.; Yang, Y.; Soh, C.K. Toward Broadband Vibration-based Energy Harvesting. *J. Intell. Mater. Syst. Struct.* **2010**, *21*, 1867–1897. [CrossRef]

43. Halim, M.A.; Park, J.Y. Theoretical modeling and analysis of mechanical impact driven and frequency up-converted piezoelectric energy harvester for low-frequency and wide-bandwidth operation. *Sens. Actuators A Phys.* **2014**, *208*, 56–65. [CrossRef]

44. Zhou, S.; Cao, J.; Inman, D.J.; Lin, J.; Li, D. Harmonic balance analysis of nonlinear tristable energy harvesters for performance enhancement. *J. Sound Vib.* **2016**, *373*, 223–235. [CrossRef]

45. Liu, H.; Lee, C.; Kobayashi, T.; Tay, C.J.; Quan, C. Investigation of a MEMS piezoelectric energy harvester system with a frequency-widened-bandwidth mechanism introduced by mechanical stoppers. *Smart Mater. Struct.* **2012**, *21*. [CrossRef]

46. Halim, M.A.; Khym, S.; Park, J.Y. Frequency up-converted wide bandwidth piezoelectric energy harvester using mechanical impact. *J. Appl. Phys.* **2013**, *114*, 044902. [CrossRef]

47. Gu, L.; Livermore, C. Impact-driven, frequency up-converting coupled vibration energy harvesting device for low frequency operation. *Smart Mater. Struct.* **2011**, *20*, 45004. [CrossRef]

48. Halim, M.A.; Park, J.Y. Piezoceramic based wideband energy harvester using impact-enhanced dynamic magnifier for low frequency vibration. *Ceram. Int.* **2015**, *41*, S702–S707. [CrossRef]

49. Vijayan, K.; Friswell, M.I.; Haddad Khodaparast, H.; Adhikari, S. Non-linear energy harvesting from coupled impacting beams. *Int. J. Mech. Sci.* **2015**, *96–97*, 101–109. [CrossRef]

50. Wickenheiser, A.M.; Garcia, E. Broadband vibration-based energy harvesting improvement through frequency up-conversion by magnetic excitation. *Smart Mater. Struct.* **2010**, *19*, 1–11. [CrossRef]

51. Tang, Q.; Li, X. Two-stage wideband energy harvester driven by multimode coupled vibration. *IEEE/ASME Trans. Mechatron.* **2015**, *20*, 115–121. [CrossRef]

52. Tang, Q.C.; Yang, Y.L.; Li, X. Bi-stable frequency up-conversion piezoelectric energy harvester driven by non-contact magnetic repulsion. *Smart Mater. Struct.* **2011**, *20*, 1–6. [CrossRef]

53. Külah, H.; Najafi, K. Energy scavenging from low-frequency vibrations by using frequency up-conversion for wireless sensor applications. *IEEE Sens. J.* **2008**, *8*, 261–268. [CrossRef]

54. Wang, C.; Zhang, Q.; Wang, W.; Feng, J. A low-frequency, wideband quad-stable energy harvester using combined nonlinearity and frequency up-conversion by cantilever-surface contact. *Mech. Syst. Signal Process.* **2018**, *112*, 305–318. [CrossRef]

55. Zhou, S.; Cao, J.; Inman, D.J.; Lin, J.; Liu, S.; Wang, Z. Broadband tristable energy harvester: Modeling and experiment verification. *Appl. Energy* **2014**, *133*, 33–39. [CrossRef]

56. Izadgoshasb, I.; Lim, Y.Y.; Tang, L.; Padilla, R.V.; Tang, Z.S.; Sedighi, M. Improving efficiency of piezoelectric based energy harvesting from human motions using double pendulum system. *Energy Convers. Manag.* **2019**, *184*, 559–570. [CrossRef]

57. Zhang, J.; Kong, L.; Zhang, L.; Li, F.; Zhou, W.; Ma, S.; Qin, L. A Novel Ropes-Driven Wideband Piezoelectric Vibration Energy Harvester. *Appl. Sci.* **2016**, *6*, 402. [CrossRef]

58. Zhang, J.; Qin, L. A tunable frequency up-conversion wideband piezoelectric vibration energy harvester for low-frequency variable environment using a novel impact- and rope-driven hybrid mechanism. *Appl. Energy* **2019**, *240*, 26–34. [CrossRef]

59. Yang, F.; Zhang, J.; Lin, M.; Ouyang, S.; Qin, L. An ultralow frequency, low intensity, and multidirectional piezoelectric vibration energy harvester using liquid as energy-capturing medium. *Appl. Phys. Lett.* **2020**, *117*, 173901. [CrossRef]

60. Chengliang, S.; Lifeng, Q.; Fang, L.; Wang, Q.M. Piezoelectric Energy Harvesting using Single Crystal Pb(Mg$_{1/3}$Nb$_{2/3}$)O$_{3-x}$PbTiO$_3$ (PMN-PT) Device. *J. Intell. Mater. Syst. Struct.* **2008**, *20*, 559–568. [CrossRef]

*micromachines*

MDPI

*Article*

# A Multi-Mode Broadband Vibration Energy Harvester Composed of Symmetrically Distributed U-Shaped Cantilever Beams

Xiaohua Huang, Cheng Zhang and Keren Dai *

School of Mechanical Engineering, Nanjing University of Science and Technology, Nanjing 210094, China; huangxiaohua@njust.edu.cn (X.H.); zhangcheng4432@njust.edu.cn (C.Z.)
* Correspondence: dkr@njust.edu.cn

**Abstract:** Using the piezoelectric effect to harvest energy from surrounding vibrations is a promising alternative solution for powering small electronic devices such as wireless sensors and portable devices. A conventional piezoelectric energy harvester (PEH) can only efficiently collect energy within a small range around the resonance frequency. To realize broadband vibration energy harvesting, the idea of multiple-degrees-of-freedom (DOF) PEH to realize multiple resonant frequencies within a certain range has been recently proposed and some preliminary research has validated its feasibility. Therefore, this paper proposed a multi-DOF wideband PEH based on the frequency interval shortening mechanism to realize five resonance frequencies close enough to each other. The PEH consists of five tip masses, two U-shaped cantilever beams and a straight beam, and tuning of the resonance frequencies is realized by specific parameter design. The electrical characteristics of the PEH are analyzed by simulation and experiment, validating that the PEH can effectively expand the operating bandwidth and collect vibration energy in the low frequency. Experimental results show that the PEH has five low-frequency resonant frequencies, which are 13, 15, 18, 21 and 24 Hz; under the action of 0.5 g acceleration, the maximum output power is 52.2, 49.4, 61.3, 39.2 and 32.1 µW, respectively. In view of the difference between the simulation and the experimental results, this paper conducted an error analysis and revealed that the material parameters and parasitic capacitance are important factors that affect the simulation results. Based on the analysis, the simulation is improved for better agreement with experiments.

**Keywords:** energy harvesting; vibration; broadband; resonant frequency



## 1. Introduction

With the development of information technologies and big data technology, wireless sensor nodes and portable electrical devices have been widely used in many fields, such as industry, healthcare and agriculture [1,2]. At present, the service life of these devices is severely limited due to the limited power and capacity loss of the battery. At the same time, the cost of replacing the battery also further restricts the use of these devices. Therefore, the question of how to harvest energy from the surrounding environment to power these devices, instead of simply using batteries, has attracted widespread attention [3–6]. There are many forms of energy in the natural environment, such as solar energy, wind energy, vibration energy and thermal energy [7–10], and the commonly used energy harvesting mechanisms include photoelectric, piezoelectric [11,12], electromagnetic [13–15], electrostatic [16] and triboelectric effects [17,18]. In particular, there are abundant vibration energy sources in the natural environment, most of which are low-frequency vibration [19]. Piezoelectric energy harvesters (PEH) are widely used for vibration energy collection due to their large power density, simple structure and ease of fabrication [20,21].

Traditional PEHs commonly have a cantilever structure, which only generates power within a limited vibration frequency range near the resonance frequency of the first mode.

When the vibration frequency of the vibration source in the surrounding environment deviates from the first resonance frequency, the output power of the energy harvester will drop sharply, resulting in a low energy harvesting efficiency and limited application scenarios.

In order to solve this problem, various methods have been proposed to broaden the operation bandwidth of the PEH. One of the solutions is to build a piezoelectric cantilever array with different sizes and resonant frequencies [22,23], which can also harvest more energy. However, the cantilever array results in a relatively larger volume, which weakens the output power density of the device. Another solution is to design an energy harvester with a single tip mass, which can generate a single resonance frequency over which maximum energy can be harvested. Therefore, the resonance frequency can be adjusted by adjusting the properties of the tip mass, such as the center of gravity of the proof mass. Based on this advantage, auto-tuning or self-tuning devices consisting of hollow masses that encapsulate moving cylinders inside were realized [24–26]. With the change in input excitation, the moving cylinder will occupy a new position inside the hollow mass, resulting in a new resonance frequency. Unfortunately, the structures of these devices are too complicated due to the additional tuning component. Moreover, nonlinear techniques, such as bi-stable systems [27,28] and Duffing oscillation [29–31], can effectively improve the frequency response with a broad bandwidth, but nonlinear vibration of the harvester would happen only when the environmental vibration intensity is strong enough to induce the stretching force in the beam of the energy harvester.

Recently, researchers have proposed the idea of multiple-degrees-of-freedom (DOF) PEH to realize multiple resonant frequencies within a certain range to achieve broadband vibration energy harvesting. Wu et al. [32] proposed a 2-DOF PEH composed of a main beam and a second beam, which is cut out inside the main beam. The two resonant frequencies can be tuned to be closer to each other by changing the tip masses. Sun and Peter [33] designed an asymmetrical U-shaped PEH to obtain two resonant frequencies which are close to each other. Further, Zhang and Hu [34] made use of the branching cantilever structure and proposed a harvester consisting of a main beam with a piezoelectric layer and several branched beams with tip masses at their free ends. The number of resonance frequencies can be increased by increasing the number of branch beams.

In this paper, a multi-mode broadband PEH is proposed. The harvester is composed of five tip masses, two U-shaped cantilever beams and a straight beam. The bandwidth of the harvester can be widened by adjusting the weight of the tip masses and the length of the cantilever beam. Simulation and experimental study are carried out to prove its validation for wideband energy harvesting. An error analysis is also carried out to reveal the impact of material parameters and parasitic capacitance on output performance, and we improved the simulation for better agreement with the experiment.

## 2. Structural Design and Fabrication

### 2.1. Structural Design

The schematic of the proposed multi-mode PEH is shown in Figure 1. As has been studied in [34], the U-shaped cantilever with two tip masses can narrow the frequency band gap between the first two resonant frequencies. Therefore, the PEH in this paper is designed to be composed of two symmetrically distributed U-shaped cantilever beams, a straight beam with three piezoelectric layers bonded on them and five tip masses. Among the five tip masses, one is attached to the free end of the straight beam, and the other four are attached to the turn-back and free end of the two U-shaped cantilever beams. The 5 resonance frequencies close to each other within a certain range are realized by specific arrangement of the multiple tip masses, and the geometric parameters are shown in Table 1.

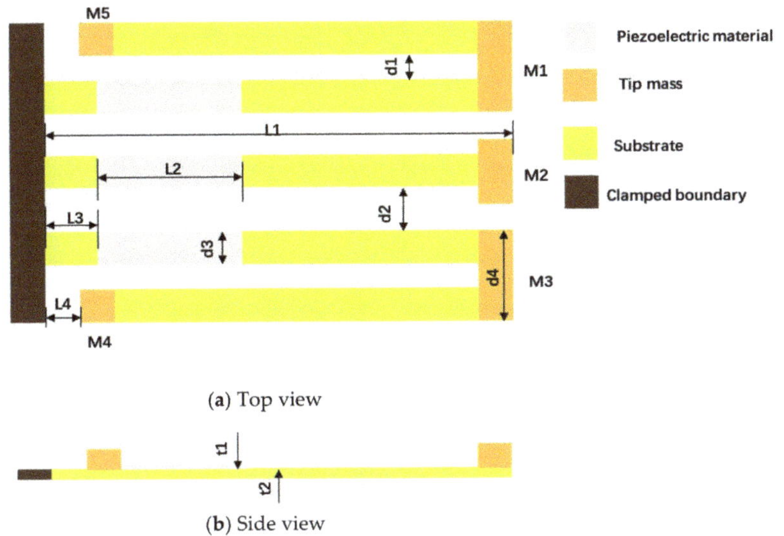



(**a**) Top view

(**b**) Side view

**Figure 1.** Schematic of the proposed PEH. (**a**) Top view; (**b**) Side view.

**Table 1.** Geometric parameters of the proposed PEH.

| Parameter | Value | Parameter | Value |
|---|---|---|---|
| L1 | 38 mm | t1 | 0.15 mm |
| L2 | 10 mm | t2 | 0.15 mm |
| L3 | 4 mm | M1 | $10 \times 4 \times 3$ mm$^3$ |
| L4 | 2 mm | M2 | $8 \times 3 \times 4$ mm$^3$ |
| d1 | 2 mm | M3 | $10 \times 4 \times 3.5$ mm$^3$ |
| d2 | 5 mm | M4 | $4 \times 3 \times 4$ mm$^3$ |
| d3 | 4 mm | M5 | $4 \times 3 \times 3$ mm$^3$ |
| d4 | 10 mm | | |

## 2.2. Fabrication

The entire fabrication process of the PEH is shown in Figure 2. The detailed processing procedure is as follows:

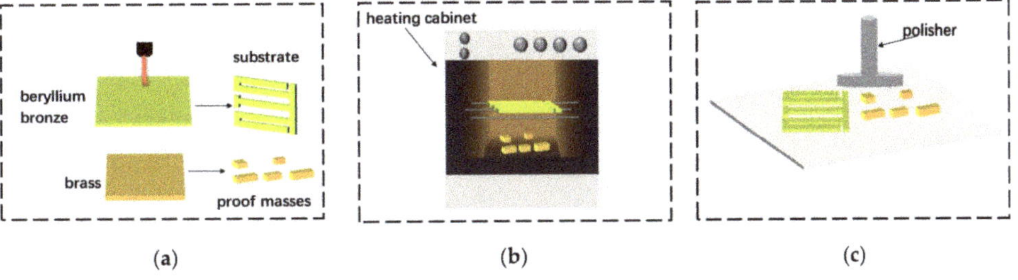



**Figure 2.** *Cont.*

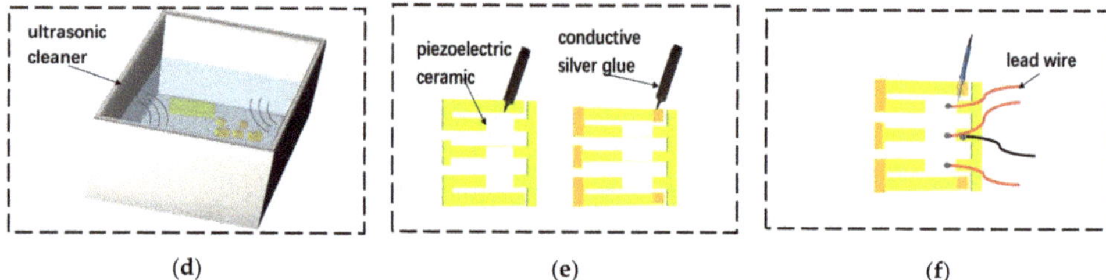


**Figure 2.** The fabrication processes of the proposed PEH. (**a**) Laser cutting; (**b**) Surface heating; (**c**) Surface polishing; (**d**) Surface cleaning; (**e**) Bonding piezoelectric ceramics and proof masses; (**f**) Wire bonding.

(a) Laser cutting. A laser cutting machine is used to cut the substrate and five tip masses according to the designed geometric dimensions.

(b) Surface heating. In order to make the surface of the metal substrate smooth, it needs to be placed in a heating box, heated at 300°C for two hours and then taken out and cooled to room temperature.

(c) Surface polishing. Place the cooled metal substrate and tip masses on a polishing machine and use the polishing machine to remove impurities and oxide layers on the surface of the substrate and five tip masses to ensure a clean and flat surface.

(d) Surface cleaning. Put the polished metal substrate and tip masses into an ultrasonic cleaning machine to clean the surface of impurities to facilitate subsequent bonding work. Put the cleaned parts on a clean glass plate and let them dry.

(e) Bonding piezoelectric ceramics and proof masses. The piezoelectric ceramic sheet and the metal substrate need to be connected with conductive silver glue to meet the requirements of mechanical connection and electrical connection. Place the piezoelectric ceramic sheet and the metal substrate on a flat operating table, use a small brush to evenly spread the conductive silver glue on the surfaces of both and then place the piezoelectric ceramic on the metal substrate and press it gently. Wipe gently with a cotton ball dipped in acetone solution to remove excess glue. Similarly, the masses are bonded. After the piezoelectric ceramics and the mass block are bonded, place the bonded structure for more than 24 h to ensure a stable bonding.

(f) Wire bonding. Place the energy harvester on the operating table, and use an electric soldering iron to weld the thin wires on the three electrodes and the upper surface of the metal substrate, respectively, as the positive and negative electrodes of the energy harvester. Then, use conductive silver glue to bond the three electrodes on the surface of the piezoelectric ceramic; the three wires on the three electrodes are connected in parallel.

### 3. Simulation and Experimental Study

In order to verify the performance of the designed PEH, a finite element simulation and experiment are carried out.

### 3.1. Simulation Study

All simulation studies on the proposed PEH are carried out in the finite element simulation software, the COMSOL Multiphysics (COMSOL, Stockholm, Swede). This is a multi-physical field coupling analysis software and provides a dedicated module that can be used to simulate piezoelectric transducers. The dimensions of the finite element model are set according to Table 1. The three piezoelectric ceramics attached to the substrate are connected in parallel; the surface of three piezoelectric ceramics and the substrate are connected to the two terminals of the load resistor, respectively. Thus, the finite element model of the proposed PEH connected with the load resistor is established. Modal analysis is carried out to determine the first six vibration modes and resonance frequencies of the

PEH. Harmonic excitation is performed in modal analysis to obtain the voltage response and the optimal matching resistance of the PEH. The simulation parameters of the proposed PEH are shown in Table 2.

**Table 2.** Material parameters of the proposed PEH.

| Property | Substrate | Piezoelectric | Tip Mass |
|---|---|---|---|
| Material | Beryllium Bronze | PZT-5H | Brass |
| Young's modulus | 128 GPa | 60.6 GPa | 110 GPa |
| Density | 8300 kg/m$^3$ | 7500 kg/m$^3$ | 8500 kg/m$^3$ |
| Piezoelectric constant (d31) | - | $-2.74 \times 10^{-10}$ C/N | - |

The first six vibration modes are shown in Figure 3; the resonance frequencies are 16.8, 18.3, 21.8, 24.7, 27.5 and 112.2 Hz. In the first five vibration modes, the deformation of the PEH is the motion of the cantilever beam along the direction of substrate thickness. In the sixth vibration mode, the cantilever beam is twisted; at the same time, its resonance frequency is much higher than the first five resonance frequencies. Therefore, only the first five resonant modes of the PEH are studied in this paper. In Figure 3, the motion of tip masses M1, M2 and M3 forms the first three resonant frequencies, respectively. The fourth and fifth resonant frequencies are caused by tip masses M4 and M5. The movement of the masses M4 and M5 drives the vibration of the masses M1 and M3, which causes the up and down vibration of the cantilever beam. Therefore, each tip mass determines the resonance frequency of the PEH. By adjusting the weight of tip masses, the resonance frequencies of the PEH can be changed, so the bandwidth of the PEH can be widened by shortening the interval between different resonance frequencies.

Modal analysis has shown that the PEH has five resonance frequencies in the low-frequency range. Frequency response analysis is performed to confirm whether the output voltage of the PEH is improved by coupling resonance frequencies. By adjusting the resistance to an extremely large value, the PEH can be considered in open-circuit condition. The frequency response of the open-circuit voltage is shown in Figure 4a. When the acceleration is 0.5 g, the proposed harvester has five voltage peaks; maximum peak voltages are 8.5, 7.6, 10.2, 8.4 and 7.1 V. It is validated that the structure of the PEH can generate multiple resonant voltages by modal coupling. According to the simulation result, as the acceleration increases, the output voltage of the PEH also increases; when the frequency deviates from the resonance frequencies, the output voltage drops rapidly. Therefore, if the frequency intervals between voltage peaks continue to expand, the operating bandwidth will decrease sharply. In this work, the frequency interval shortening mechanism is used to increase operation bandwidth.

The output power is an important indicator for evaluating the performance of a PEH. The PEH is equivalent to a power supply to provide energy to the load, and the load can be simply viewed as a resistor. In order to obtain the maximum output power from the PEH and improve the working efficiency of the PEH, the optimal load resistance value of the PEH need to be determined based on the principle of impedance matching. Therefore, the relationship between the output power and the load resistance is studied. Although the optimal resistance value at each frequency will be different, the output power near the resonance frequency is the largest and the most noteworthy. Therefore, the optimal resistance at the resonance frequency is the most valuable to study. Since the output voltage of the PEH is the largest at the third resonance frequency, this article seeks the optimal resistance value at the third-order resonance frequency.

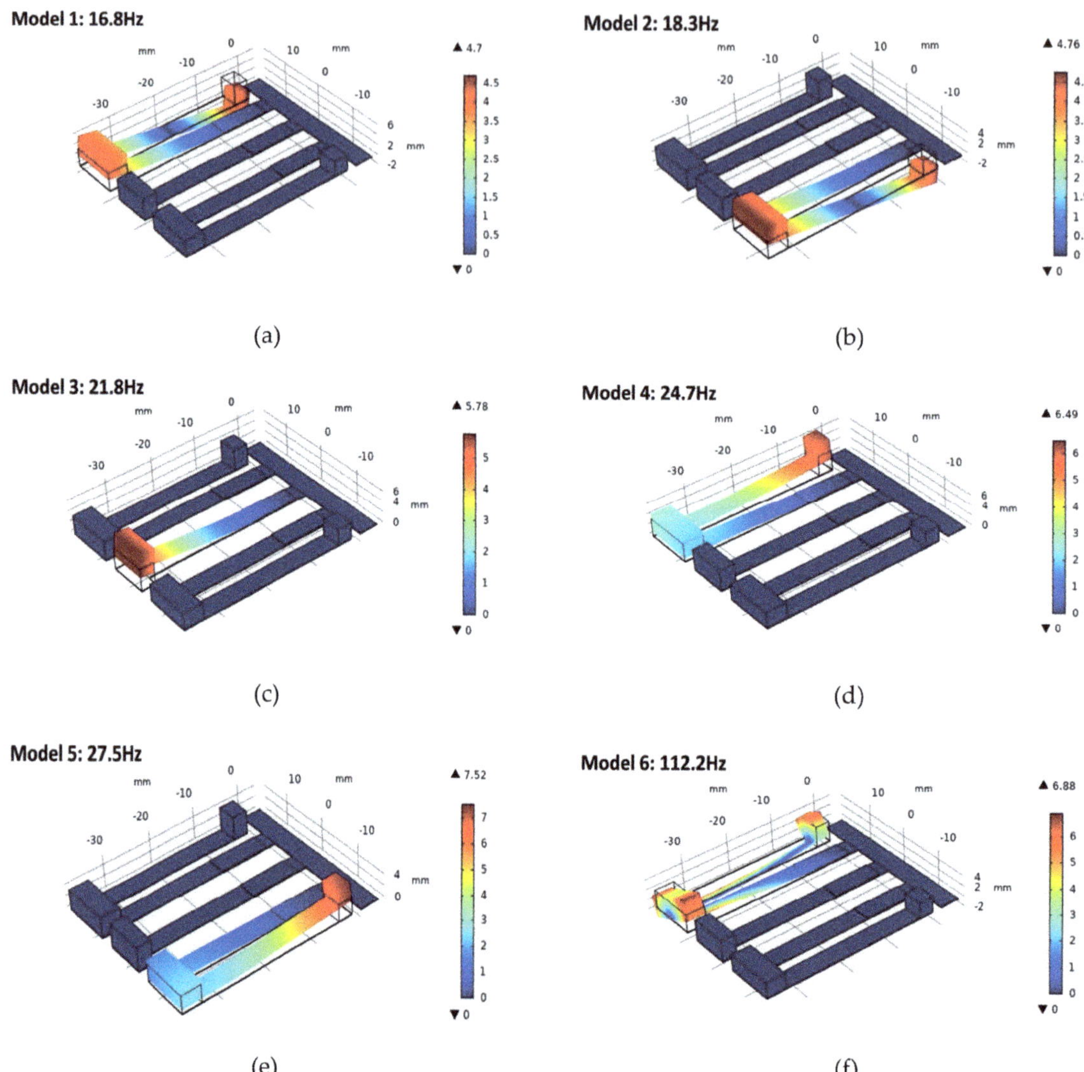



(a)

(b)

(c)

(d)

(e)

(f)

**Figure 3.** (**a–f**) Finite element simulation of multiple mode shapes of the proposed structure.

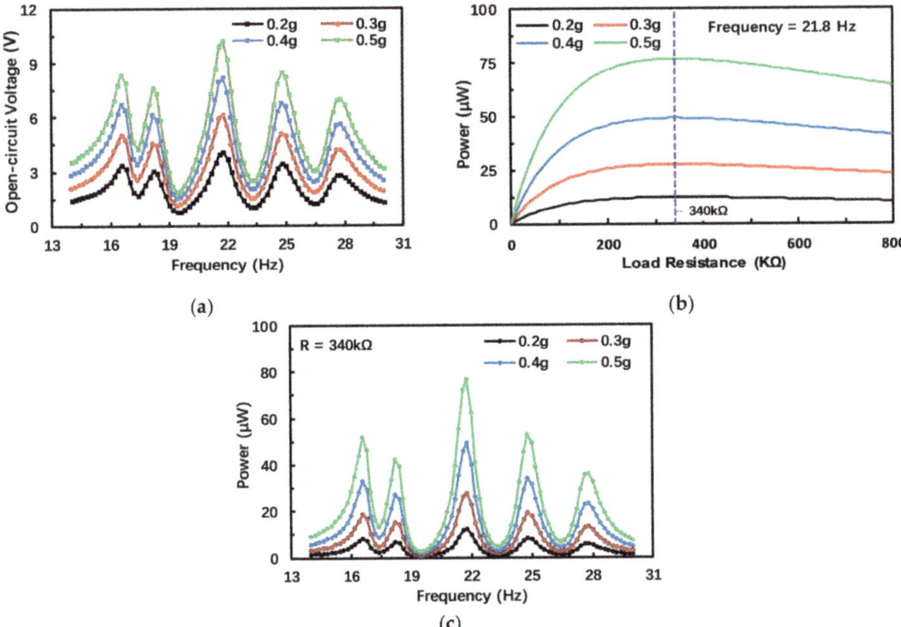

**Figure 4.** Simulation results of the output characteristics of the PEH. (**a**) Simulation results of frequency response of open-circuit voltage; (**b**) Simulation results of output power against load resistance under different acceleration amplitudes; (**c**) Simulation results of frequency response of output power.

The relationship between output power and load resistance is shown in Figure 4b (the acceleration amplitude changes from 0.2 to 0.5 g with an interval of 0.1 g). In the simulation study, the variable resistor value increases from 0 to 800 kΩ in intervals of 10 kΩ. According to the simulation result in Figure 4, as the resistance increases, the output power of the harvester first increases and then decreases. Therefore, the optimal load resistance for the harvester is 340 kΩ, and only in this case can the PEH output maximum power and improve energy harvesting efficiency.

In order to analyze the output power of the PEH at different frequencies, frequency response analysis of the output power is carried out, and simulation results are shown in Figure 4c. The output power of the PEH with the optimal resistor is similar to the open-circuit voltage, with five peaks at similar frequency points in the low-frequency range. At the third resonance frequency, the output power can reach up to 76.88 μW, when the acceleration is 0.5 g. At the resonance frequency, the output power is the maximum; when the working frequency deviates from the resonance frequency, the output power of the PEH will decrease. Due to the small interval between the resonance frequencies, the average output power of the PEH within a certain range of the resonance frequencies could maintain a relatively high level to satisfy practical application.

Since the geometry sizes of different PEH are different, it is necessary to calculate the power density of the PEH to evaluate its performance. Figure 5 shows the comparison of the power density between this work and an M-shaped folded cantilever PEH [4], when two harvesters are under the action of sinusoidal signals of different frequencies. Consider the operating bandwidth of the two harvesters: the frequency of sinusoidal signals increases from 12 to 28 Hz at an interval of 1 Hz. The acceleration signal actually applied to the PEH is

$$y = \sum_{k=12}^{28} 0.05g \sin(2\pi kt) \tag{1}$$

where $g$ is the acceleration of gravity and $t$ is the vibration time. The simulation time set in this paper is one second. The volume of piezoelectric material is used to calculate the power density. As shown in Figure 5, the maximum power density of the PEH designed in this paper is 1.2732 $\mu$Wmm$^3$, and the maximum power density in [4] is 0.5112 $\mu$Wmm$^3$; the maximum power density of this paper is nearly 2.49 times this value. During the whole simulation period, the power density of the proposed PEH is larger than that in [4] for most of the time, and the average power density is 0.1733 and 0.1230 $\mu$Wmm$^3$, respectively. The simulation result confirms that the proposed harvester has higher output power both in terms of maximum power density, average power density and, most of the time, instantaneous power density under a complex environment.

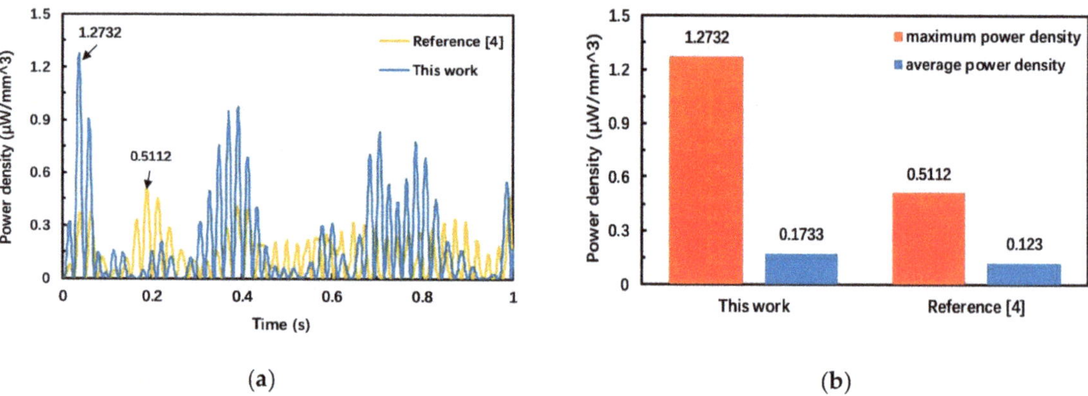


**Figure 5.** Comparison of output power density of two harvesters. (**a**) Instantaneous output power density during simulation; (**b**) Comparison of maximum and average power density.

### 3.2. Experimental Study

#### 3.2.1. Experimental Setup

The feasibility of the structure has been verified through simulation, and we next need to build a test platform to test the real object of the PEH. The schematic of the testing system for the proposed PEH is shown in Figure 6. By adjusting the frequency of the harmonic excitation signal of the function generator (YB1602, Lvyang, Yangzhou, China) and the output current of the power amplifier (SA-P050, Shiao, Wuxi, China), the shaker (SA-JZ050, Shiao, Wuxi, China) can generate vibration signals of different frequencies and amplitudes. In the experiment, the excitation frequency is manually swept from 10 to 26 Hz. During this sweeping procedure, the amplitude of the vibration signal generated by the shaker is measured by the accelerometer (HWT901B, Wit, Shenzhen, China) as a feedback loop. Due to the large internal resistance of the PEH, the open-circuit voltage generated by the harvester is measured with an electrometer (Keithley6514, Tektronix, Cleveland, OH, USA), and the measured voltage data are sent to a computer via a data acquisition card (NI USB-6363, National Instruments, Austin, TX, USA).

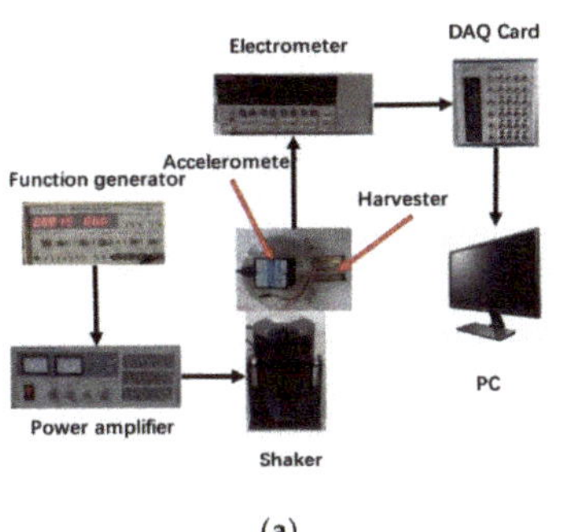


(a)

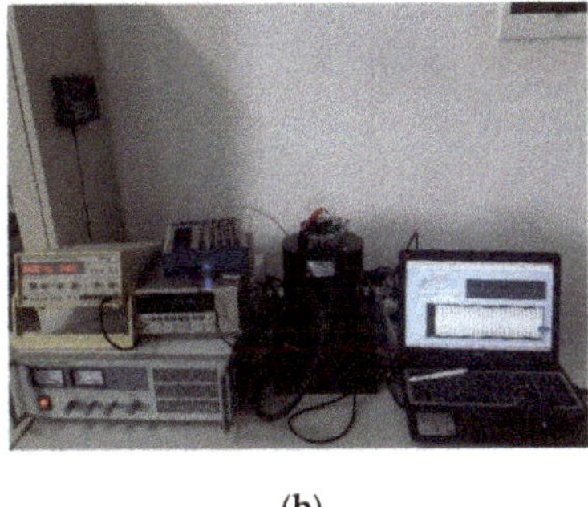

(b)

**Figure 6.** (**a**) Schematic of experimental setup for the PEH testing; (**b**) The experimental scene diagram of the PEH test.

### 3.2.2. Experimental Results

The experiment measured the frequency response of the open-circuit voltage of the PEH under different acceleration amplitudes. The measurement results are shown in Figure 7a. When the acceleration is 0.5 g, the PEH has five voltage peaks in the range of 11–26 Hz, with peak values of 6.4, 6.2, 6.9, 5.1 and 6.4 V, which further confirms the feasibility of the structure of the PEH. The frequency corresponding to the peak voltage is the resonance frequency of the PEH, so the first five resonant frequencies are 13, 15, 18, 21 and 24 Hz. Similarly, the output voltage of the PEH increases with the acceleration amplitude, and the maximum voltage is obtained at the resonance frequency. The output voltage decreases away from the resonance frequency. Compared with the simulation results, the open-circuit voltage and first five resonance frequencies measured by the experiment are smaller than the simulation values.

In order to find the optimal load resistance of the PEH, we studied the relationship between the output power of the PEH and the load resistance. In our experiment, a variable resistor ranging from 10 to 600 kΩ in intervals of 20 kΩ is applied to study the performance with different resistances. The optimal resistance value also appears at the third resonance frequency. Figure 7b shows the relationship between the output power and the load resistance at the third resonance frequency, which is 18 Hz. The output power first increases and then decreases with the increase in resistance. When the load resistance is 260 kΩ, the output power can reach the maximum, approximately 61.3 μW, at an acceleration of 0.5 g, so the optimal resistor value is 260 kΩ for this harvester. Since the output voltage of the PEH is lower than the simulated value, its output power is also lower than the simulated value. Nowadays, many small sensing systems can operate normally with low power consumption. For example, the power consumption of a temperature sensing system in [35] and a pressure measurement microsystem in [36] is only 71 nW and 120 μW (for the whole microsystem), respectively. In addition, these microsystems usually operate at intervals, and the power consumption can be further satisfied with the aid of energy storage and management systems such as super capacitors. Therefore, the proposed PEH can supply these devices directly or through energy storage devices successfully for practical application.

Figure 7c shows the frequency response of the output power of the PEH under different acceleration amplitudes and a load resistance of 260 kΩ. The experimental results and the simulation results display the same changing trend. The output power is the largest at the resonance frequency; when the ambient frequency deviates from the resonance frequency, the output power will decrease.

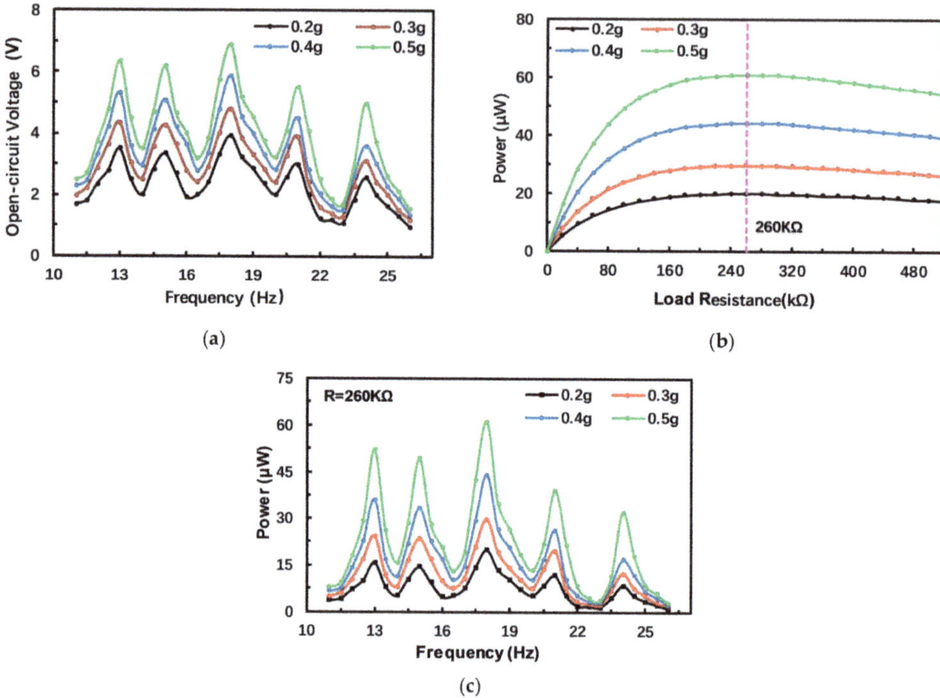

**Figure 7.** Experimental results of output characteristics of the PEH. (**a**) Measured frequency response of open-circuit voltage; (**b**) Measured results of output power against load resistance under different acceleration amplitudes; (**c**) Measured results of frequency response of output power.

### 4. Error Analysis

Simulation and experimental results validate that the PEH does have five resonant frequencies in the low-frequency range, and the variation trend of the open-circuit voltage and output power of the PEH is the same in the simulation and experiment. However, there are some deviations in the values of open-circuit voltage and resonance frequency, and the resonance frequency in the experiment is generally around 3 Hz lower than the simulated value. To further study the source of these deviations, the effects of the Young's modulus of the substrate on the resonance frequency of the proposed PEH are studied. The influence of the Young's modulus of the substrate on the resonance frequency is shown in Figure 8. As the Young's modulus decreases, the first five resonance frequencies also decrease.

In addition to the Young's modulus, parasitic capacitance will also have an important impact on the output performance of the PEH [37,38]. The PEH can be regarded as an equivalent circuit network composed of infinite parallel branches composed of inductance, capacitance, resistance, ideal voltage source and ideal transformer, where each parallel branch represents a certain order of vibration mode. The parasitic capacitance usually comes from the internal capacitances of the piezoelectric transducer; at the same time, there are parasitic capacitances in the surrounding environment, such as the external wires, the metal in the environment and capacitances in the measuring equipment. The influence of parasitic capacitance in the environment also needs to be added to the equivalent circuit model. As shown in Figure 9a, the PEH can be represented as an ideal voltage source $V_r$ in series with a capacitor $C_r$, a resistance $R_r$, an inductance $L_r$ and finally in parallel with an ideal transformer $N_r$. The capacitor $C_p$ is the parasitic capacitance and is connected in parallel with the PEH in the equivalent circuit model.

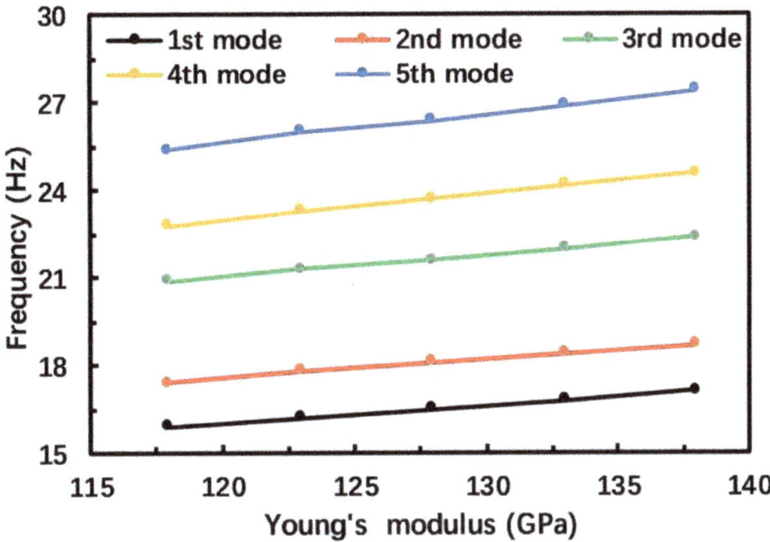

**Figure 8.** The influence of the Young's modulus of the substrate on the resonance frequencies.

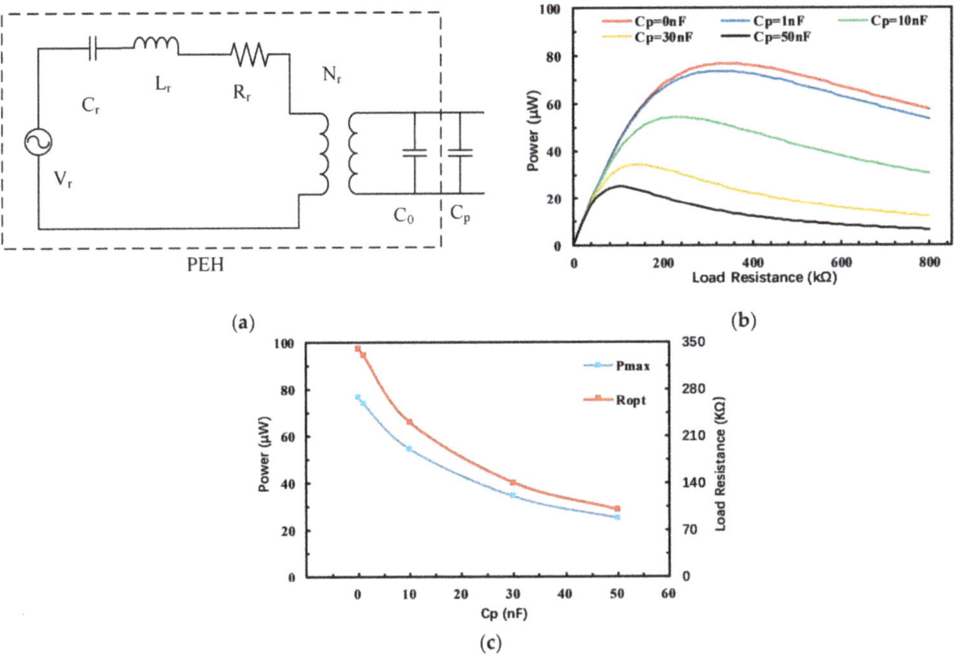

**Figure 9.** The influence of the parasitic capacitance on output characteristics. (**a**) The equivalent circuit model for the PEH with parasitic capacitances; (**b**) The relationship between output power and resistance for different parasitic capacitances; (**c**) The relationship between maximum output power, optimal matching resistance and parasitic capacitance.

The output characteristics of the PEH for different parasitic capacitances are calculated by simulation, and the results are shown in the Figure 9b. It can be clearly seen that the parasitic capacitance has a strong influence on the output characteristics of the PEH. When the external resistance is small, the output power difference of the PEH is very small, but as the load resistance increases, the larger the parasitic capacitance of the PEH, and the more the output power drops. The relationships between the maximum output power, optimal matching resistance and parasitic capacitance are shown in Figure 9c. As the parasitic capacitance increases, the maximum output power and optimal matching resistance will both decrease rapidly. When the parasitic capacitance increases from 0 to 50 nF, the maximum output power of the PEH is reduced by 62.3% and the optimal matching resistance is changed from 340 to 100 kΩ.

Considering the influence of the above factors, the modified parameters (Young's modulus: 118 GPa, parasitic capacitance: 15 nF) are taken for simulation analysis. It can be seen from Figure 10 that the simulation results using the modified parameters essentially coincide with the experimental results, and the errors are significantly reduced in terms of both output power and resonance frequencies.

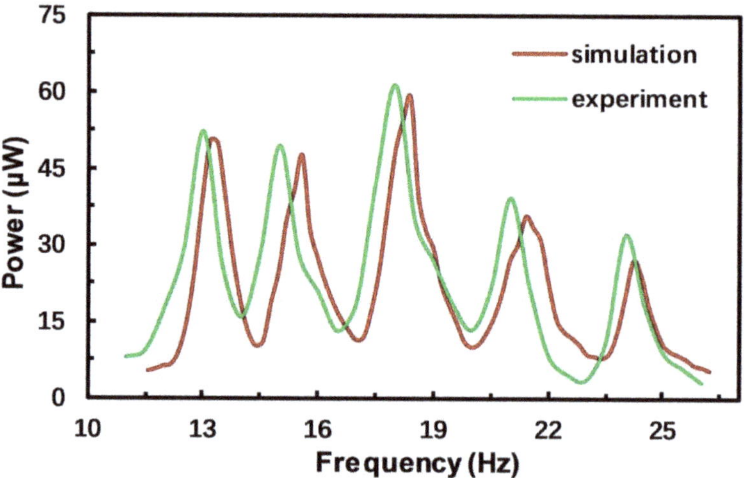


**Figure 10.** Comparison of simulation and experimental results of frequency response.

### 5. Conclusions

This article proposed a multi-mode broadband PEH composed of symmetrically distributed U-shaped cantilever beams, a straight beam and five tip masses. A piece of piezoelectric ceramic is bonded to each beam. Through structural optimization design, the number of voltage peaks of the PEH device in the low-frequency range is further increased. The experiment and simulation are conducted to validate the feasibility of the proposed structure. The finite element simulation and experiment results show that the PEH can indeed generate five voltage peaks in the range of 10–30 Hz. The power density of the PEH proposed in this article is approximately 1.8 times that of an asymmetric M-shaped cantilever PEH due to its broad working frequency. A theoretical analysis of the error between the idealized modeling, simulation and the non-idealized experimental test results for this type of PEH device reveals the influence of multiple non-ideal factors. The Young's modulus of the substrate will affect the resonance frequencies of the PEH, and the presence of parasitic capacitance will reduce the maximum output power, frequencies of maximum power and optimal matching resistance. With an adjustment of these parameters, the simulation is improved for better agreement with the experiment.

The proposed PEH realized vibration energy harvesting in a wide frequency range and achieved high power density, indicating its great potential for powering wireless sensor nodes and portable electronic devices in practical application scenarios. The error analysis also provides guidance for the simulation of PEH devices.

**Author Contributions:** Conceptualization: X.H., C.Z. and K.D.; methodology—fabrication: X.H.; data curation and analysis: C.Z.; writing—original draft preparation: X.H., C.Z. and K.D.; writing—review and editing: X.H. and K.D.; All authors have read and agreed to the published version of the manuscript.

**Funding:** This research received no external funding.

**Data Availability Statement:** Data are contained within the article.

**Conflicts of Interest:** The authors declare no conflict of interest.

# References

1. Zou, H.X.; Zhao, L.C.; Gao, Q.H.; Zuo, L.; Liu, F.R.; Tan, T.; Zhang, W.M. Mechanical modulations for enhancing energy harvesting: Principles, methods and applications. *Appl. Energy* **2019**, *255*, 113871. [CrossRef]
2. Dong, L.; Closson, A.B.; Jin, C.; Trase, I.; Chen, Z.; Zhang, J.X. Vibration-Energy-Harvesting System: Transduction Mechanisms, Frequency Tuning Techniques, and Biomechanical Applications. *Adv. Mater. Technol.* **2019**, *4*, 1900177. [CrossRef]
3. Oh, Y.; Kwon, D.S.; Eun, Y.; Kim, W.; Kim, M.O.; Ko, H.J.; Kim, J. Flexible energy harvester with piezoelectric and thermoelectric hybrid mechanisms for sustainable harvesting. *Int. J. Precis. Eng. Manuf. -Green Technol.* **2019**, *6*, 691–698. [CrossRef]
4. Wu, M.; Ou, Y.; Mao, H.; Li, Z.; Liu, R.; Ming, A.; Ou, W. Multi-resonant wideband energy harvester based on a folded asymmetric M-shaped cantilever. *AIP Adv.* **2015**, *5*, 077149. [CrossRef]
5. Zhang, C.; Dai, K.; Liu, D.; Yi, F.; Wang, X.; Zhu, L.; You, Z. Ultralow Quiescent Power-Consumption Wake-Up Technology Based on the Bionic Triboelectric Nanogenerator. *Adv. Sci.* **2020**, *7*, 2000254. [CrossRef]
6. Qi, N.; Dai, K.; Yi, F.; Wang, X.; You, Z.; Zhao, J. An adaptive energy management strategy to extend battery lifetime of solar powered wireless sensor nodes. *IEEE Access* **2019**, *7*, 88289–88300. [CrossRef]
7. Zhou, G.; Huang, L.; Li, W.; Zhu, Z. Harvesting ambient environmental energy for wireless sensor networks: A survey. *J. Sens.* **2014**, *2014*, 1–20. [CrossRef]
8. Hwang, W.; Kim, K.B.; Cho, J.Y.; Yang, C.H.; Kim, J.H.; Song, G.J.; Sung, T.H. Watts-level road-compatible piezoelectric energy harvester for a self-powered temperature monitoring system on an actual roadway. *Appl. Energy* **2019**, *243*, 313–320. [CrossRef]
9. Nguyen, M.S.; Yoon, Y.J.; Kim, P. Enhanced broadband performance of magnetically coupled 2-DOF bistable energy harvester with secondary intrawell resonances. *Int. J. Precis. Eng. Manuf. -Green Technol.* **2019**, *6*, 521–530. [CrossRef]
10. Zhong, Y.; Zhao, H.; Guo, Y.; Rui, P.; Shi, S.; Zhang, W.; Wang, Z.L. An Easily Assembled Electromagnetic-Triboelectric Hybrid Nanogenerator Driven by Magnetic Coupling for Fluid Energy Harvesting and Self-Powered Flow Monitoring in a Smart Home/City. *Adv. Mater. Technol.* **2019**, *4*, 1900741. [CrossRef]
11. Yang, Z.; Zu, J. Toward harvesting vibration energy from multiple directions by a nonlinear compressive-mode piezoelectric transducer. *IEEE ASME Trans Mechatron* **2015**, *21*, 1787–1791. [CrossRef]
12. Zhao, H.; Wei, X.; Zhong, Y.; Wang, P. A Direction Self-Tuning Two-Dimensional Piezoelectric Vibration Energy Harvester. *Sensors* **2020**, *20*, 77. [CrossRef]
13. Liu, H.; Hou, C.; Lin, J.; Li, Y.; Shi, Q.; Chen, T.; Lee, C. A non-resonant rotational electromagnetic energy harvester for low-frequency and irregular human motion. *Appl. Phys. Lett.* **2018**, *113*, 203901. [CrossRef]
14. Fan, K.; Cai, M.; Liu, H.; Zhang, Y. Capturing energy from ultra-low frequency vibrations and human motion through a monostable electromagnetic energy harvester. *Energy* **2019**, *169*, 356–368. [CrossRef]
15. Chen, J.; Wang, Y. A dual electromagnetic array with intrinsic frequency up-conversion for broadband vibrational energy harvesting. *Appl. Phys. Lett.* **2019**, *114*, 053902. [CrossRef]
16. Gao, C.; Gao, S.; Liu, H.; Jin, L.; Lu, J.; Li, P. Optimization for output power and band width in out-of-plane vibration energy harvesters employing electrets theoretically, numerically and experimentally. *Microsyst. Technol.* **2017**, *23*, 5759–5769. [CrossRef]
17. Xu, M.; Zhao, T.; Wang, C.; Zhang, S.L.; Li, Z.; Pan, X.; Wang, Z.L. High power density tower-like triboelectric nanogenerator for harvesting arbitrary directional water wave energy. *ACS Nano* **2019**, *13*, 1932–1939. [CrossRef] [PubMed]
18. Wang, P.; Pan, L.; Wang, J.; Xu, M.; Dai, G.; Zou, H.; Wang, Z.L. An ultra-low-friction triboelectric–electromagnetic hybrid nanogenerator for rotation energy harvesting and self-powered wind speed sensor. *ACS Nano* **2018**, *12*, 9433–9440. [CrossRef] [PubMed]
19. Kim, J.E.; Lee, S.; Kim, Y.Y. Mathematical model development, experimental validation and design parameter study of a folded two-degree-of-freedom piezoelectric vibration energy harvester. *Int. J. Precis. Eng. Manuf. Technol.* **2019**, *6*, 893–906. [CrossRef]
20. Wang, P.; Liu, R.; Ding, W.; Zhang, P.; Pan, L.; Dai, G.; Wang, Z.L. Complementary Electromagnetic-Triboelectric Active Sensor for Detecting Multiple Mechanical Triggering. *Adv. Funct. Mater.* **2018**, *28*, 1705808. [CrossRef]
21. Yang, J.; Yue, X.; Wen, Y.; Li, P.; Yu, Q.; Bai, X. Design and analysis of a 2D broadband vibration energy harvester for wireless sensors. *Sens. Actuators A* **2014**, *205*, 47–52. [CrossRef]

22. Mažeika, D.; Čeponis, A.; Yang, Y. Multifrequency piezoelectric energy harvester based on polygon-shaped cantilever array. *Noise Control* **2018**, *2018*, 1–11. [CrossRef]
23. Priya, S. Criterion for material selection in design of bulk piezoelectric energy harvesters. *IEEE Trans. Sonics Ultrason.* **2010**, *57*, 2610–2612. [CrossRef] [PubMed]
24. Somkuwar, R.; Chandwani, J.; Deshmukh, R. Wideband auto-tunable vibration energy harvester using change in centre of gravity. *Microsyst. Technol.* **2018**, *24*, 3033–3044. [CrossRef]
25. Chandwani, J.; Somkuwar, R.; Deshmukh, R. Multi-band piezoelectric vibration energy harvester for low-frequency applications. *Microsyst. Technol.* **2019**, *25*, 3867–3877. [CrossRef]
26. Chandwani, J.; Somkuwar, R.; Deshmukh, R. Experimental study on band merging of non-linear multi-band piezoelectric energy harvester into single broadband using magnetic coupling. *Microsyst. Technol.* **2020**, *26*, 657–671. [CrossRef]
27. Li, X.; Li, Z.; Huang, H.; Wu, Z.; Huang, Z.; Mao, H.; Cao, Y. Broadband spring-connected bi-stable piezoelectric vibration energy harvester with variable potential barrier. *Results Phys.* **2020**, *18*, 103173. [CrossRef]
28. Yang, W.; Towfighian, S. A hybrid nonlinear vibration energy harvester. *Mech. Syst. Signal Process.* **2017**, *90*, 317–333. [CrossRef]
29. Hajati, A.; Kim, S.G. Ultra-wide bandwidth piezoelectric energy harvesting. *Appl. Phys. Lett.* **2011**, *99*, 083105. [CrossRef]
30. Erturk, A.; Hoffmann, J.; Inman, D.J. A piezomagnetoelastic structure for broadband vibration energy harvesting. *Appl. Phys. Lett.* **2009**, *94*, 254102. [CrossRef]
31. Hajati, A.; Bathurst, S.P.; Lee, H.J.; Kim, S.G. Design and fabrication of a nonlinear resonator for ultra wide-bandwidth energy harvesting applications. In *2011 IEEE 24th International Conference on Micro Electro Mechanical Systems*; IEEE: New York, NY, USA, 2011.
32. Wu, H.; Tang, L.; Yang, Y.; Soh, C.K. A compact 2 degree-of-freedom energy harvester with cut-out cantilever beam. *Jpn. J. Appl. Phys.* **2012**, *51*, 040211. [CrossRef]
33. Sun, S.; Peter, W.T. Modeling of a horizontal asymmetric U-shaped vibration-based piezoelectric energy harvester (U-VPEH). *Mech. Syst. Signal Process.* **2019**, *114*, 467–485. [CrossRef]
34. Zhang, G.; Hu, J. A branched beam-based vibration energy harvester. *J. Electron. Mater.* **2014**, *43*, 3912–3921. [CrossRef]
35. Lee, Y. A review of recent research on mm-scale sensor systems. In *2015 International SoC Design Conference (ISOCC)*; IEEE: New York, NY, USA, 2015; pp. 87–88.
36. Ziaie, B.; Najafi, K. An implantable microsystem for tonometric blood pressure measurement. *Biomed. Microdevices* **2001**, *3*, 285–292. [CrossRef]
37. Dai, K.; Wang, X.; Niu, S.; Yi, F.; Yin, Y.; Chen, L.; You, Z. Simulation and structure optimization of triboelectric nanogenerators considering the effects of parasitic capacitance. *Nano Res.* **2017**, *10*, 157–171. [CrossRef]
38. Kanda, K.; Saito, T.; Iga, Y.; Higuchi, K.; Maenaka, K. Influence of parasitic capacitance on output voltage for series-connected thin-film piezoelectric devices. *Sensors* **2012**, *12*, 16673–16684. [CrossRef]

 *micromachines*

*Article*

# Power Density Improvement of Piezoelectric Energy Harvesters via a Novel Hybridization Scheme with Electromagnetic Transduction

Zhongjie Li [1,2], Chuanfu Xin [1], Yan Peng [1,3,*], Min Wang [1,3,*], Jun Luo [1,3], Shaorong Xie [1,3] and Huayan Pu [1,3]

1   School of Mechatronic Engineering and Automation, Shanghai University, Shanghai 200444, China; lizhongjie@shu.edu.cn (Z.L.); xcfshu@shu.edu.cn (C.X.); luojun@shu.edu.cn (J.L.); srxie@shu.edu.cn (S.X.); phygood_2001@shu.edu.cn (H.P.)
2   Shanghai Institute of Intelligent Science and Technology, Tongji University, Shanghai 200092, China
3   Engineering Research Center of Unmanned Intelligent Marine Equipment, Ministry of Education, 99 Shangda Rd., Shanghai 200444, China
*   Correspondence: pengyan@shu.edu.cn (Y.P.); xmwangmin@shu.edu.cn (M.W.)

**Abstract:** A novel hybridization scheme is proposed with electromagnetic transduction to improve the power density of piezoelectric energy harvester (PEH) in this paper. Based on the basic cantilever piezoelectric energy harvester (BC-PEH) composed of a mass block, a piezoelectric patch, and a cantilever beam, we replaced the mass block by a magnet array and added a coil array to form the hybrid energy harvester. To enhance the output power of the electromagnetic energy harvester (EMEH), we utilized an alternating magnet array. Then, to compare the power density of the hybrid harvester and BC-PEH, the experiments of output power were conducted. According to the experimental results, the power densities of the hybrid harvester and BC-PEH are, respectively, 3.53 mW/cm$^3$ and 5.14 μW/cm$^3$ under the conditions of 18.6 Hz and 0.3 g. Therefore, the power density of the hybrid harvester is 686 times as high as that of the BC-PEH, which verified the power density improvement of PEH via a hybridization scheme with EMEH. Additionally, the hybrid harvester exhibits better performance for charging capacitors, such as charging a 2.2 mF capacitor to 8 V within 17 s. It is of great significance to further develop self-powered devices.

**Keywords:** piezoelectric; electromagnetic; hybrid energy harvester; power density improvement



**Citation:** Li, Z.; Xin, C.; Peng, Y.; Wang, M.; Luo, J.; Xie, S.; Pu, H. Power Density Improvement of Piezoelectric Energy Harvesters via a Novel Hybridization Scheme with Electromagnetic Transduction. *Micromachines* **2021**, *12*, 803. https://doi.org/10.3390/mi12070803

Academic Editors: Qiongfeng Shi and Huicong Liu

Received: 18 May 2021
Accepted: 29 June 2021
Published: 7 July 2021

**Publisher's Note:** MDPI stays neutral with regard to jurisdictional claims in published maps and institutional affiliations.



## 1. Introduction

The extensive use of smart sensor devices (such as marine environment monitoring wireless sensors, wildfire detectors, etc.), which are usually powered by batteries, leads to a large amount of power consumption. However, due to the limited lifespan of batteries, it severely restricts the continuous operation of smart sensor devices and replacing batteries regularly can cause high costs. To solve these problems, developing self-powered devices by using some approaches to harvest ambient energy, for instance, solar energy [1–3], wind [4–6], tidal energy [7], thermal energy [8,9], mechanical vibration [10,11], and pyroelectric energy [12,13], is a feasible program. The commonly used energy harvesting mechanism is piezoelectric [14–17].

The piezoelectric energy conversion mechanism has been used to extract kinetic energy because of the high power density, simple operation mechanism, and design flexibility [18]. The basic configuration of PEH is composed of a mass block, a piezoelectric patch, and a cantilever beam. Based on the basic configuration, many novel designs were proposed. Zhang et al. [19] designed a multi-impact harvester with superior performance to extract energy under the low-frequency vibration. Iman et al. [20] collected energy from human motion by using a harvester and powered electronic devices. Based on impact vibration, a low-frequency PEH, assembled with two rigid generating beams and a compliant driving beam, was presented by Gu, which can achieve an average power of 1.53 mW under the conditions of 20.1 Hz and 0.4 g [21]. Besides, some original designs of PEHs were proposed, including a cantilever PEH with a revolute joint [22], a press-button type PEH [23], a PEH with a sandwich structure [24], a T-shaped PEH with internal resonance [25], a truss-based compressive-mode PEH [26], a PEH based on the piezoelectric stack [27,28], and a U-shaped bi-directional PEH [29].

Recently, piezoelectric-electromagnetic hybrid harvesters have attracted gravitational attention [30]. Iqbal et al. [31] presented a hybrid harvester, which contains the PEH and EMEH, to collect low-frequency vibration energy from walking motion. The harvester can generate the power of 51 μW and 36 μW from EMEH and PEH, respectively. Iqbal et al. [32] designed a multimodal hybrid bridge harvester. The power of 2214.32 μW and 155.7 μW were, respectively, generated from its electromagnetic and piezoelectric portions. Edwards et al. [33] introduced a novel low-frequency vibration energy harvester and conducted a series of simulations and experiments. Based on the experimental results, the harvester can, respectively, achieve the average power of 46.2 μW and 3.6 μW from electromagnetic and piezoelectric transducers at 5 Hz. Pyo et al. [34] investigated a hybrid harvester by using frequency up-conversion, which can extract energy from an extremely low-frequency mechanical motion and generate peak power of 6.03 mW and 1.35 mW from electromagnetic and piezoelectric portions, respectively.

In the above work, researchers mostly use a magnet or a simple combination of several magnets to form the hybrid energy harvester, which can result in small output power of EMEHs. Additionally, Li et al. [35] have proven that an alternating magnet array, which can cause abrupt magnetic flux density changes, can improve the output power of EMEHs. Therefore, we proposed a novel hybrid scheme with an alternating magnet array to improve the power density of PEHs.

In this paper, we presented a hybrid harvester based on the BC-PEH, which displayed excellent performance under the weak excitation. The key contributions include: first, designing a novel hybrid scheme to improve the power density of BC-PEH; second, implementing a simulation to testify the abrupt magnetic flux density (MFD) changes caused by an alternating magnet array; third, conducting experiments of open-circuit voltage, frequency sweep, output power, and charging capacitors to compare the output performance of the hybrid harvester and the BC-PEH; and finally, further demonstrating the high power density of the hybrid harvester by charging millifarad level capacitors.

## 2. Configuration and Simulation

Based on the BC-PEH composed of a mass block, a piezoelectric patch, and a cantilever beam, we replaced the mass block by a magnet array and added a coil array to form the hybrid energy harvester (PEH source and EMEH source). Due to the magnetic damping, the deformation of the piezoelectric material becomes small, so the output power of the PEH decreases. However, the EMEH generates power in the reciprocating motion of the magnet array. Therefore, to compare the total output power of the hybrid harvester and the BC-PEH, we made a prototype and conducted experiments of open-circuit voltage, frequency sweep, output power, and charging capacitors.

The configuration of the hybrid harvester was illustrated in Figure 1a. It is composed of a base, a cantilever beam, a coil bracket, and a magnet frame. In Figure 1a, one end of the cantilever beam and the coil bracket are fixed to the base, and the magnet frame is connected with the other end of the cantilever beam. As shown in Figure 1b, the magnet array contains two magnets, which are installed in the magnet frame, and a proof mass is constituted of the magnet array and magnet frame. As shown in Figure 1c, the coil array placed on either side of the magnet array, including four coils, is mounted in the coil bracket. An electromagnetic energy conversion mechanism is constructed by the coil array and proof mass. Additionally, a piezoelectric patch (as shown in Figure 1a) acts as a piezoelectric energy conversion mechanism with the cantilever beam and proof mass.

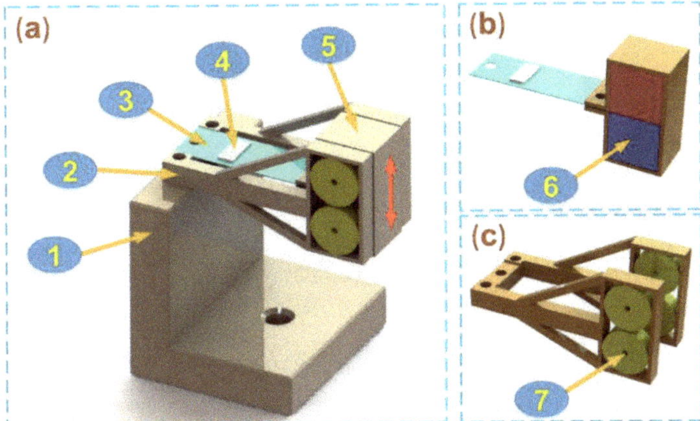

**Figure 1.** The configuration of the hybrid harvester. The components are, respectively: 1-base; 2-coil bracket; 3-cantilever beam; 4-piezoelectric patch; 5-magnet frame; 6-magnet; and 7-coil. (**a**) The configuration. (**b**) The magnet array. (**c**) The coil array.

When the hybrid harvester is applied excitations, the proof mass is forced to reciprocate in the indicated direction in Figure 1a, which causes the cantilever beam to bend. Then, the bending induces the piezoelectric patch to deform and leads to strain generation inside the material. Therefore, it yields the voltage according to the direct piezoelectric effect. At the same time, relative movement occurs between the proof mass and the coil array. Therefore, the coils generate an induced electromotive force ($E_{mf}$), which can be gauged by using Equation (1).

$$E_{mf} = -N\frac{d(BS)}{d(t)} \tag{1}$$

where, $N$, $B$, $S$, and $t$ are the turns of the coil array, the MFD, the area of the coil, and the time, respectively. In the reciprocating motion of the proof mass, the cantilever beam firstly goes from the initial position to the point of maximum upward displacement, then starts moving down, then reaches the position of maximum downward displacement, and then returns to the initial position. As shown in Figure 2a,c, the cantilever beam is at the initial position and does not deform, so the output voltage of the PEH is zero, which is found in Figure 2e (P1 and P3). However, the proof mass gets the largest movement speed at this moment, so the maximum instantaneous voltage is obtained from the EMEH according to Equation (1), as shown in Figure 2f (E1 and E3). Figure 2b,d show the cantilever beam is at the positions of maximum upward displacement and maximum downward displacement. At this time, the piezoelectric patch achieves the maximum deformation, so the PEH yields the maximum output voltage, as shown in Figure 2e (P2 and P4). Besides, the movement speed of the proof mass is zero, so the EMEH does not generate the output voltage, as shown in Figure 2f (E2 and E4).

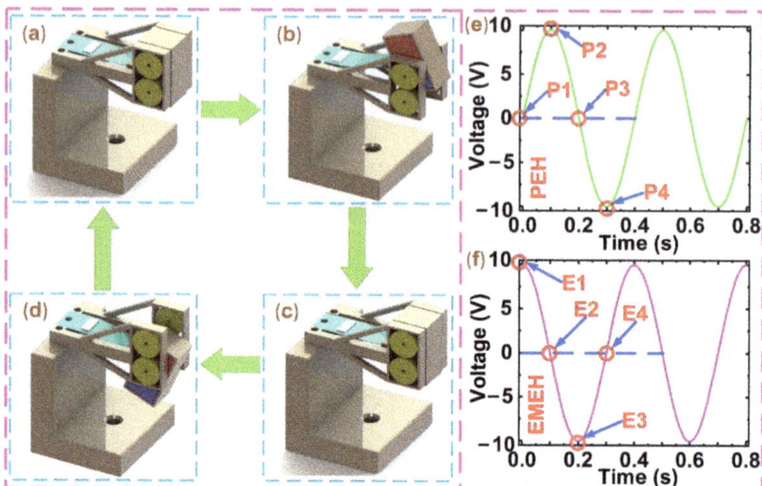



**Figure 2.** The positions of the cantilever beam in the reciprocating motion of the proof mass. (**a**) The initial position. (**b**) The position of maximum upward displacement. (**c**) The initial position. (**d**) The position of maximum downward displacement. (**e**) The simulated voltage of the PEH. (**f**) The simulated voltage of the EMEH.

According to Equation (1), we can enhance the output voltages of EMEHs by increasing the change rate of the MFD. Moreover, the alternating magnet array can cause abrupt magnetic flux density changes. Therefore, we utilize the way of alternating arrangement of magnetic poles in the scheme. According to the model shown in Figure 3a, the simulation was conducted to demonstrate this phenomenon mentioned above by using COMSOL Multiphysics 5.4 (Sweden). First, based on Figure 3a, we constructed two cube magnets with a length of 12 mm in COMSOL. The magnetic flux (1 T) directions of the two magnets are opposite along the Z-axis of the coordinate system. Then, the MFD distribution was depicted in the Z direction of the model. Figure 3b displays a step change in MFD between two continuous magnets. Namely, the MFD immediately decreases from the peak value to the minimal value, which verifies the phenomenon above. Moreover, the maximum value of the MFD is on the polar surface of magnets, so the distance between coils and magnets should be as small as possible.

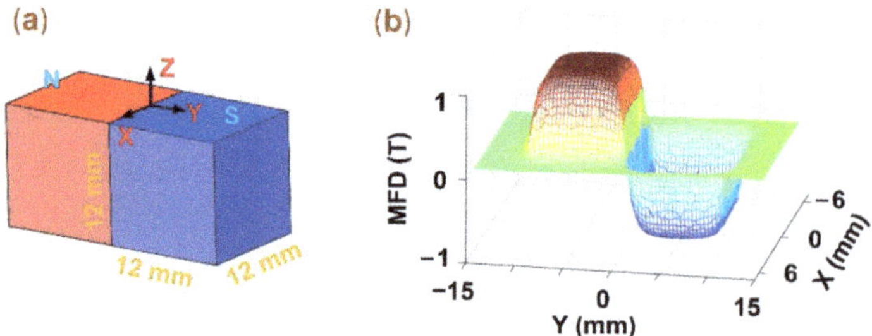



**Figure 3.** The simulation of the MFD. (**a**) the simulation model; (**b**) the MFD distribution in the Z direction of the model.

## 3. Experiment and Discussion

### 3.1. Prototype Fabrication and Experiment Setup

According to the configuration in Figure 1a, a prototype was fabricated and the experiments of open-circuit voltage, frequency sweep, output power, and charging capacitors were conducted to examine the output performance of the prototype and compare the outputs with/without the hybridization scheme. The prototype and experiment conditions are shown in Figure 4. The coil bracket and one end of the cantilever beam were fixed to the base with screws, and the magnet frame was joined to the other end of the cantilever beam with screws and nuts. The base, cantilever beam, coil bracket, and magnet frame of the hybrid energy harvester were all made of copper. Furthermore, the distance between the coil and the magnet was set to 0.5 mm.

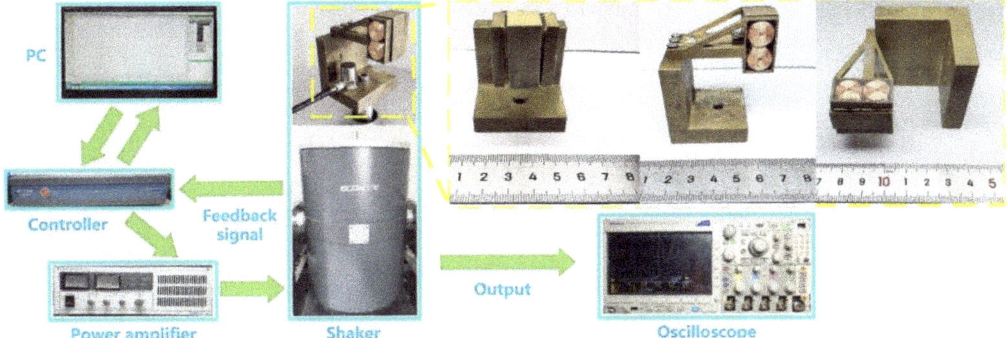


**Figure 4.** The prototype and experiment conditions.

The types of the shaker, accelerometer, controller, power amplifier, and oscilloscope, which were used in the experiment, are Econ E-JZK-50, Econ EA-YD-181, Econ VT-9002, Econ E5874A (ECON, Kunshan, China) and Tektronix mdo3024 (Tektronix, OK, USA), respectively. The prototype was fixed on the shaker with a screw, and the accelerometer was fixed on the base by using tape. Then, the relevant parameters were set on the PC, and the experiments of open-circuit voltage, frequency sweep, output power, and charging capacitors were conducted. The control signal was generated by the controller after receiving the instructions from the PC. When the power amplifier got the control signal from the controller, the shaker was driven. Then, the feedback signal was transmitted to the controller by the accelerometer, so the hybrid energy harvester works under a constant acceleration. The output voltages of the hybrid energy harvester were all measured by a digital oscilloscope. The detailed material properties and geometric parameters of the prototype are shown in Table 1.

**Table 1.** Material properties and the geometric parameters of the prototype.

| Description | | Value |
|---|---|---|
| Prototype | Dimensions (mm$^3$) | $49 \times 23 \times 26$ |
| Cantilever beam | Dimensions (mm$^3$) | $35 \times 12 \times 0.5$ |
| Magnet array | Number | 1 |
| | Number of magnets | 2 |
| Magnet | Dimensions (mm$^3$) | $12 \times 12 \times 12$ |
| | Magnet grade | N52 |
| | Material | NdFeB |
| Coil array | Number | 2 |
| | Number of coils | 4 |
| Coil | Outside dimension (mm$^3$) | $12 \times 4.7$ |
| | Inside dimension (mm$^3$) | $1.8 \times 4.7$ |
| | Number of turns | 1860 |
| | Resistance ($\Omega$) | 85.3 |
| | Wire diameter (mm) | 0.1 |

Table 1. *Cont.*

| Description | | Value |
|---|---|---|
| Piezoelectric patch | Piezoelectric material | PZT-5H |
| | Dimensions (mm$^3$) | $7.5 \times 3.5 \times 0.5$ |

### 3.2. Experiments of Open-Circuit Voltage

The open-circuit voltages of the hybrid energy harvester (PEH source and EMEH source) were measured under the harmonic excitation of different frequencies (14.6 Hz, 16.6 Hz, 18.6 Hz, 20.6 Hz, and 22.6 Hz) and constant acceleration (0.3 g). The open-circuit voltages of the PEH and EMEH are shown in Figure 5a,b.

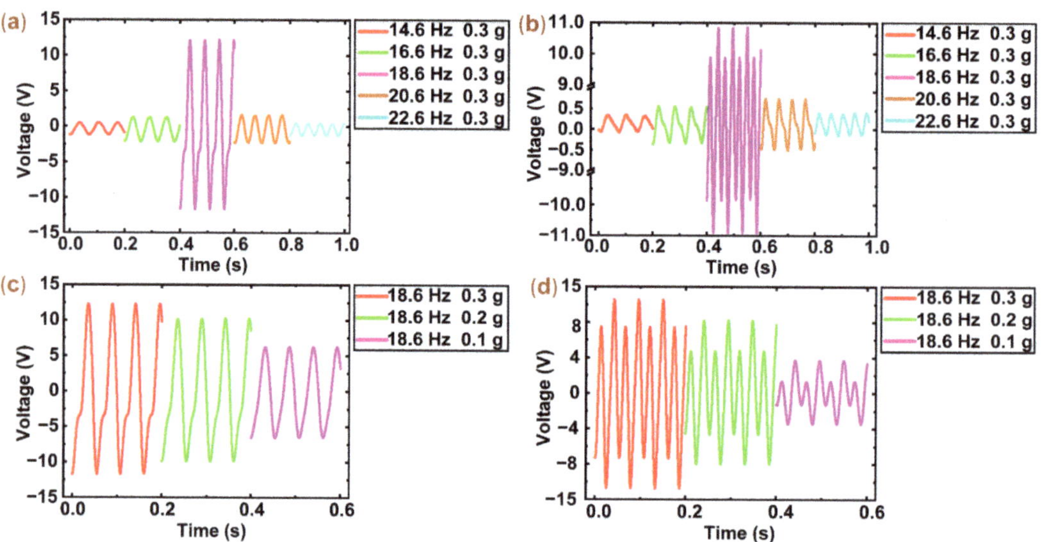

**Figure 5.** The output voltages of the PEH and EMEH under different frequencies and acceleration. (**a**,**b**) The output voltages of the PEH and EMEH under different frequencies (14.6 Hz, 16.6 Hz, 18.6 Hz, 20.6 Hz, and 22.6 Hz) and constant acceleration (0.3 g). (**c**,**d**) The output voltages of the PEH and EMEH under constant frequency (18.6 Hz) and different acceleration (0.1 g, 0.2 g, and 0.3 g).

In Figure 5a, a peak-to-peak open-circuit voltage of the PEH is 2.0 V under the excitation frequency of 14.6 Hz. The voltage increases with the rise of excitation frequency and reaches a maximum of 25.3 V at 18.6 Hz. Then, the voltage decreases with the increase of excitation frequency and the voltage is 1.9 V at 22.6 Hz. As shown in Figure 5b, the voltage change of the EMEH is similar to that of the PEH. The voltage is 0.58 V at 14.6 Hz. As the excitation frequency increases to 18.6 Hz, the voltage reaches a peak of 21.9 V. Then, when the excitation frequency is 22.6 Hz, the voltage decreases to 0.74 V.

The voltage changes of the PEH and EMEH are the same behavior pattern. From 14.6 Hz to 18.6 Hz, the voltage continually increases. The voltage changes of the frequencies from 14.6 Hz to 16.6 Hz are less obvious than that of the frequencies from 16.6 Hz to 18.6 Hz. From 18.6 Hz to 22.6 Hz, the voltages unremittingly decrease. The voltage changes of the frequencies from 18.6 Hz to 20.6 Hz are more observable than that of the frequencies from 20.6 Hz to 22.6 Hz. When the excitation frequency increases to near the resonant frequency, the maximum deformation of the piezoelectric patch and the peak rate of magnetic flux density change is achieved. Therefore, the voltages of the PEH and EMEH display an obvious change.

To further study open-circuit voltages of the PEH and EMEH, we measured the voltages of the PEH and EMEH under constant frequency (18.6 Hz) and different excitation acceleration (0.1 g, 0.2 g, 0.3 g). In Figure 5c, the voltage of the PEH is 13.01 V at 0.1 g. As the excitation acceleration rises, the voltage increases. When the excitation acceleration is 0.2 g and 0.3 g, the voltages are 20.65 V and 25.73 V, respectively. As shown in Figure 5d, the voltage trend of the EMEH and the PEH is similar. When the excitation acceleration is 0.1 g, 0.2 g, and 0.3 g, the voltages are 7.34 V, 16.47 V, and 21.97 V, respectively.

In Figure 5, the open-circuit voltage waveform of the PEH and EMEH is not a standard harmonic. They, respectively, show resembling sawtooth waves and alternating sine waves of different amplitude in the resonant region. At resonance, the deformation of the cantilever beam is large and non-linear, which results in the strain of the piezoelectric patch and the displacement change of the magnet array being non-linear. Therefore, the voltage waveform of the PEH and EMEH displays a nonstandard sine waveform.

### 3.3. Experiments of Frequency Sweep

Afterward, to compare the output voltages of the BC-PEH and the hybrid harvester, the frequency-sweep experiments were conducted. In these experiments, the sweep rate and frequency domain are 0.1 Hz/s and (15 Hz, 23 Hz), respectively. Based on the different acceleration (0.1 g, 0.2 g, and 0.3 g), the open-circuit voltages of the BC-PEH and hybrid harvester (PEH source and EMEH source) were measured. The experimental results are shown in Figure 6.

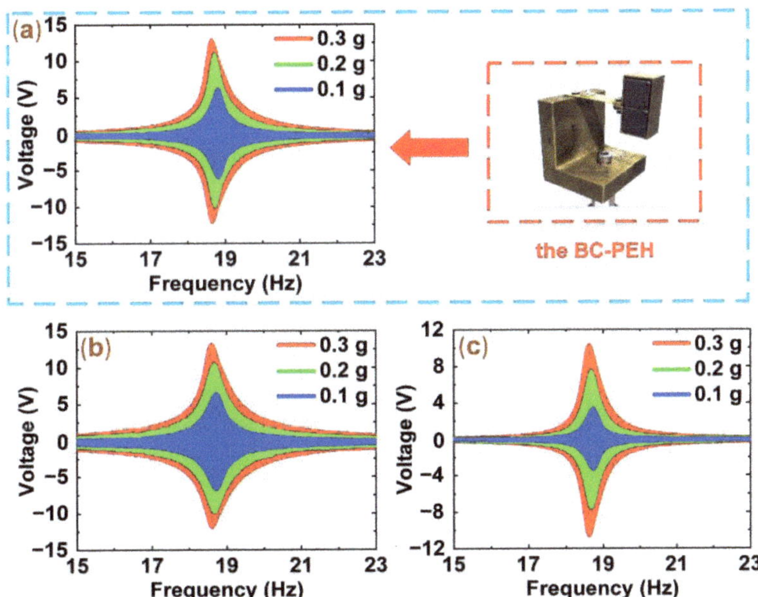


**Figure 6.** The output voltages of the BC-PEH and the hybrid harvester under different acceleration (0.1 g, 0.2 g, and 0.3 g). (**a**) The output voltages of the BC-PEH. (**b**) The output voltages of the PEH. (**c**) The output voltages of the EMEH.

Figure 6a–c display the resonant frequency (18.6 Hz) of the prototype. The voltages of the BC-PEH and hybrid harvester vary as the excitation acceleration increases and reach a maximum at 18.6 Hz. The voltages of the PEH are 25.7 V, 20.6 V, and 13.1 V under the resonant frequency and acceleration of 0.3 g, 0.2 g, and 0.1 g, respectively. The behavior patterns of the voltage changes of the BC-PEH and EMEH are similar to that of the PEH. The voltages of the BC-PEH are 25.2 V, 21.1 V, and 12.3 V under the acceleration of 0.3 g,

0.2 g, and 0.1 g, respectively. Similarly, the voltage changes of 21.9 V, 16.5 V, and 7.4 V for EMEH are obtained under the acceleration of 0.3 g, 0.2 g, and 0.1 g, respectively.

In the sweep experiments, the current value of the EMEH is approximately 0, so the damping of the BC-PEH and hybrid harvester mainly comes from the mechanical damping. Therefore, as shown in Figure 6a,b, the voltage values of the BC-PEH and PEH are nearly the same.

### 3.4. Experiments of Output Power

To compare the output performance of the BC-PEH and hybrid harvester, the output power experiments were carried out under the conditions of 0.3 g and 18.6 Hz. In these experiments, we recorded the power, voltages, and currents of the BC-PEH and hybrid harvester (PEH source and EMEH source). The results are shown in Figure 7.

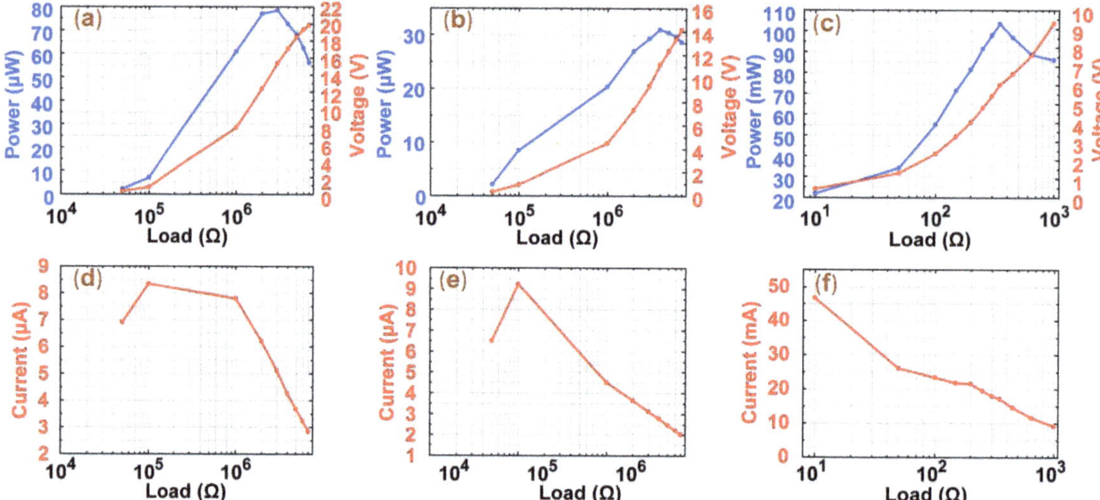

**Figure 7.** The output power, voltages, and currents of the BC-PEH and hybrid harvester. (**a–c**) The peak-to-peak output power and voltages of the BC-PEH, PEH, and EMEH. (**d–f**) The peak-to-peak output currents of the BC-PEH, PEH, and EMEH.

Figure 7b,e show the changes in the power, voltages, and currents of the PEH under different external electric resistances. The power of the PEH gradually increases and reaches a maximum of 31.08 μW and then decreases. The voltages of the PEH continue to enlarge until it approaches the open-circuit voltage as the external electric resistances increase. In addition, the currents of the PEH slightly rise and then continuously decline. Figure 7a,c,d,f show the changes of the power, voltages, and currents of the BC-PEH and EMEH, which are similar to that of the PEH. The maximum power of the BC-PEH and EMEH is 78.5 μW and 103.5 mW, respectively. Compared to the power density of 5.14 μW/cm$^3$ of the BC-PEH, the power density of 3.53 mW/cm$^3$ of the hybrid harvester is 686 times as high as the BC-PEH. Besides, it should be mentioned that the peak power of the BC-PEH is higher than the PEH, which is attributed to the electromagnetic damping.

To further explore the ability of the hybrid harvester as a power source to supply power to an external load, the average power ($P_{avg}$) of the hybrid harvester was calculated. The voltage waveform of the BC-PEH and hybrid harvester is a nonstandard sine waveform, so we use Equation (2) to obtain the average power of the BC-PEH and hybrid harvester. The average power of the hybrid harvester is approximately 7.98 mW and the BC-PEH is

4.38 µW. Therefore, the average power of the hybrid harvester is 1821 times as high as the BC-PEH.

$$P_{avg} = \int_0^t U^2(t)/R_m dt/T \tag{2}$$

where $U(t)$, $R_m$, $P_{avg}$, and $T$ are the instantaneous voltage, matching impedance, average power, and full-time of an excitation, respectively.

In Table 2, we displayed a performance comparison for the proposed prototype and other hybrid harvesters with different configurations.

**Table 2.** The performance comparison of hybrid harvesters.

| Ref. | Dimensions (cm$^3$) | Frequency (Hz) | Excitation (Speed or Acceleration) | Power (mW) | Power Density (mW/cm$^3$) |
|---|---|---|---|---|---|
| [36] | 105 × 30 × 20 | 17 | 0.4 g | 15.82 | 0.251 |
| [37] | 3.85 × 3.4 × 3.7 | / | 4–6 km/h | 0.55 | 1.14 × 10$^{-3}$ |
| [38] | 46.8 | 51 | 0.5 g | 1.67 | 0.036 |
| [39] | 1.84 | 2 | / | 0.0298 | 0.01619 |
| [40] | 5 | 23.3 | 0.4 g | 2.26 | 0.452 |
| [41] | 10 × 4 × 0.1 | 33.5 | 0.3 g | 3.32 | 0.83 |
| [42] | 4 × 1.5 × 4 | 113.5 | 0.6 g | 3.54 | 0.1475 |
| [43] | 19.2 | 5.2 | 2 g | 1.2288 | 0.064 |
| [44] | 4 × 4 × 1 | / | 1 m/s | 14.0135 | 0.876 |
| [45] | 6 × 0.7 × 2 | 16.8 | 0.5 g | 3.12 | 0.371 |
| This paper | 4.9 × 2.3 × 2.6 | 18.6 | 0.3 g | 103.51 | 3.53 |

The power density of the prototype can reach 3.53 mW/cm$^3$ under a weak acceleration (0.3 g) in this paper, which is 14 times, 7 times, and 4 times as high as Refs [34,38,42]. Therefore, the output performance of the prototype in this paper is superior.

*3.5. Experiments of Charging Capacitors*

To further compare the output power of the BC-PEH and hybrid harvester, we conducted charging experiments. We firstly designed an experimental circuit, which contains two harvesters and rectifiers, as shown in Figure 8a. Since the open-circuit voltage magnitudes of the PEH and EMEH are at the same level, we chose to connect the PEH and EMEH in parallel. Two rectifiers were, respectively, connected to the ends of the PEH and EMEH and converted the negative current signals of the PEH and EMEH into positive ones.

In the charging experiments, we selected six capacitors: 470 µF, 2.2 mF, 3.3 mF, 4.7 mF, 6.8 mF, and 10 mF, and set the experimental conditions as follows: a constant acceleration of 0.3 g, a constant frequency of 18.6 Hz. We used the BC-PEH and the hybrid harvester to conduct comparative charging experiments. Before experiments, we fully discharged the capacitors. For the capacitor of 470 µF, it was only charged to 1.95 V within 120 s by using the BC-PEH, as shown in Figure 8d. When we used the hybrid harvester to charge the capacitors of 2.2 mF, 3.3 mF, 4.7 mF, 6.8 mF, and 10 mF, they were, respectively, charged to 8 V, 7.2 V, 6.3 V, 4.9 V, and 4.1 V within 17 s, as shown in Figure 8c. Compared to the BC-PEH, the hybrid harvester can charge a larger capacitor to a higher voltage in a shorter period. The hybrid harvester exhibits excellent charge performance.

It should be mentioned that the average power ($P_a$) for charging capacitors can be calculated by Equation (3):

$$P_a = C_p(U_2{}^2 - U_1{}^2)/2\Delta t \tag{3}$$

where $P_a$, $C_P$, $U_1$, $U_2$ and $\Delta t$ represent the average charging power, capacity, initial voltage, final voltage, and charging time, respectively. To compare the power of the BC-PEH and the hybrid harvester, we made a histogram of the power, as shown in Figure 8b. When the capacitor of 10 mF was charged, the power of the hybrid harvester reached 4.94 mW. Compared to the power of 0.0147 mW for charging the capacitor of 470 µF by using the BC-PEH, the average charging power of the hybrid harvester is 336 times higher than that of the BC-PEH. In addition, to store more energy at the same time, it is necessary to choose a capacitor with a large capacity.

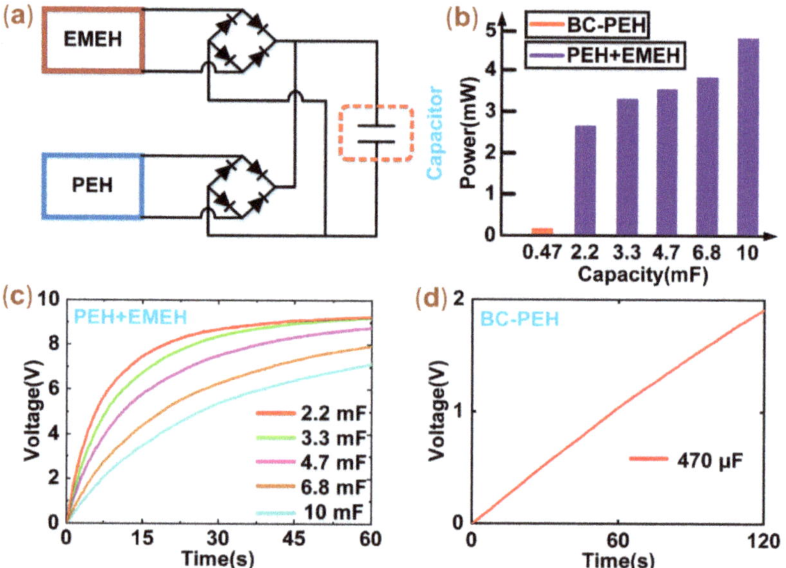


**Figure 8.** The experimental circuit and results of charging capacitors. (**a**) The circuit of charging capacitors. (**b**) The histogram of the average charging power. (**c**) The results of charging capacitors by using the hybrid harvester. (**d**) The result of charging capacitor by using the BC-PEH.

## 4. Conclusions

This paper proposed a novel hybridization scheme with electromagnetic transduction to enhance the power density of PEHs. The hybrid energy harvester was designed based on the BC-PEH. To compare the power density of the BC-PEH and the hybrid energy harvester, we built a prototype and conducted experiments of open-circuit voltage, frequency sweep, output power, and charging capacitors. According to the experimental results, the key conclusions of this paper are as follows:

1. The EMEH can yield a high voltage of 21.9 V under a weak acceleration of 0.3 g by using an alternating magnet array, which can result in abrupt magnetic flux density changes.
2. Comparing the peak power of the BC-PEH and hybrid harvester, the output power (103.53 mW) of the hybrid harvester is 1318 times as high as the output power (78.5 µW) of the BC-PEH.
3. Comparing the power densities and average power of the BC-PEH and hybrid harvester, the power density and average power of the hybrid harvester are, respectively, 686 times and 1821 times higher than that of the BC-PEH.
4. The hybrid harvester also displays excellent charging performance because of the high output power. According to the experimental results, the average charging power of the hybrid harvester is 336 times higher than that of the BC-PEH.

The hybrid energy harvester shows a better energy capture performance, which verifies that the power density improvement of PEHs can use a hybridization scheme with electromagnetic transduction and also displays its great potential to successfully power low-power electronic components.

**Author Contributions:** Z.L. conceived the idea and conducted the experiments. C.X. wrote the original draft. Z.L. and M.W. reviewed and revised the paper. Y.P. provided financial support. J.L. and S.X. helped fabricated the prototype. H.P. helped with data processing. All authors have read and agreed to the published version of the manuscript.

**Funding:** This work was funded and supported by National Natural Science Foundation of China (No.: 61773254; No.: 62001281) and Shanghai Sailing Program (No.: 20YF1412700).

**Institutional Review Board Statement:** Not applicable.

**Informed Consent Statement:** Not applicable.

**Data Availability Statement:** The data supporting the findings of this paper is available from the corresponding authors on request.

**Conflicts of Interest:** The authors declare no conflict of interest.

# References

1. Guo, C.X.; Guai, G.H.; Li, C.M. Graphene Based Materials: Enhancing Solar Energy Harvesting. *Adv. Energy Mater.* **2011**, *1*, 448–452. [CrossRef]
2. Lin, G.-J.; Wang, H.-P.; Lien, D.-H.; Fu, P.-H.; Chang, H.-C.; Ho, C.-H.; Lin, C.-A.; Lai, K.-Y.; He, J.-H. A broadband and omnidirectional light-harvesting scheme employing nanospheres on Si solar cells. *Nano Energy* **2014**, *6*, 36–43. [CrossRef]
3. Ma, W.; Li, X.; Lu, H.; Zhang, M.; Yang, X.; Zhang, T.; Wu, L.; Cao, G.; Song, W. A flexible self-charged power panel for harvesting and storing solar and mechanical energy. *Nano Energy* **2019**, *65*, 104082. [CrossRef]
4. Wu, N.; Wang, Q.; Xie, X. Wind energy harvesting with a piezoelectric harvester. *Smart Mater. Struct.* **2013**, *22*, 095023. [CrossRef]
5. Orrego, S.; Shoele, K.; Ruas, A.; Doran, K.; Caggiano, B.; Mittal, R.; Kang, S.H. Harvesting ambient wind energy with an inverted piezoelectric flag. *Appl. Energy* **2017**, *194*, 212–222. [CrossRef]
6. Nabavi, S.; Zhang, L. Portable Wind Energy Harvesters for Low-Power Applications: A Survey. *Sensors* **2016**, *16*, 1101. [CrossRef]
7. Bryden, I.; Grinsted, T.; Melville, G. Assessing the potential of a simple channel to deliver useful energy. *Appl. Ocean Res.* **2004**, *26*, 198–204. [CrossRef]
8. Wang, Q.; Bowen, C.R.; Lewis, R.; Chen, J.; Lei, W.; Zhang, H.; Li, M.-Y.; Jiang, S. Hexagonal boron nitride nanosheets doped pyroelectric ceramic composite for high-performance thermal energy harvesting. *Nano Energy* **2019**, *60*, 144–152. [CrossRef]
9. Kishore, R.; Priya, S.J. A Review on Low-Grade Thermal Energy Harvesting: Materials, Methods and Devices. *Materials* **2018**, *11*, 1433. [CrossRef] [PubMed]
10. Wei, C.; Jing, X. A comprehensive review on vibration energy harvesting: Modelling and realization. *Renew. Sustain. Energy Rev.* **2017**, *74*, 1–18. [CrossRef]
11. Won, S.S.; Seo, H.; Kawahara, M.; Glinsek, S.; Lee, J.; Kim, Y.; Jeong, C.K.; Kingon, A.I.; Kim, S.-H. Flexible vibrational energy harvesting devices using strain-engineered perovskite piezoelectric thin films. *Nano Energy* **2019**, *55*, 182–192. [CrossRef]
12. Battista, L.; Mecozzi, L.; Coppola, S.; Vespini, V.; Grilli, S.; Ferraro, P. Graphene and carbon black nano-composite polymer absorbers for a pyro-electric solar energy harvesting device based on LiNbO$_3$ crystals. *Appl. Energy* **2014**, *136*, 357–362. [CrossRef]
13. Kang, M.; Yeatman, E.M. Coupling of piezo- and pyro-electric effects in miniature thermal energy harvesters. *Appl. Energy* **2020**, *262*, 114496. [CrossRef]
14. Fan, K.; Chang, J.; Pedrycz, W.; Liu, Z.; Zhu, Y. A nonlinear piezoelectric energy harvester for various mechanical motions. *Appl. Phys. Lett.* **2015**, *106*, 223902. [CrossRef]
15. Cao, J.; Wang, W.; Zhou, S.; Inman, D.; Lin, J. Nonlinear time-varying potential bistable energy harvesting from human motion. *Appl. Phys. Lett.* **2015**, *107*. [CrossRef]
16. Kuang, Y.; Zhu, M. Characterisation of a knee-joint energy harvester powering a wireless communication sensing node. *Smart Mater. Struct.* **2016**, *25*, 055013. [CrossRef]
17. Fan, K.; Liu, Z.; Liu, H.; Wang, L.; Zhu, Y.; Yu, B. Scavenging energy from human walking through a shoe-mounted piezoelectric harvester. *Appl. Phys. Lett.* **2017**, *110*, 143902. [CrossRef]
18. Wu, N.; Bao, B.; Wang, Q. Review on engineering structural designs for efficient piezoelectric energy harvesting to obtain high power output. *Eng. Struct.* **2021**, *235*, 112068. [CrossRef]
19. Zhang, Y.; Cai, C.; Zhang, W. Experimental study of a multi-impact energy harvester under low frequency excitations. *Smart Mater. Struct.* **2014**, *23*, 055002. [CrossRef]
20. Izadgoshasb, I.; Lim, Y.Y.; Lake, N.; Tang, L.; Padilla, R.V.; Kashiwao, T. Optimizing orientation of piezoelectric cantilever beam for harvesting energy from human walking. *Energy Convers. Manag.* **2018**, *161*, 66–73. [CrossRef]
21. Gu, L. Low-frequency piezoelectric energy harvesting prototype suitable for the MEMS implementation. *Microelectron. J.* **2011**, *42*, 277–282. [CrossRef]

22. Li, Z.; Naguib, H.E. Effect of revolute joint mechanism on the performance of cantilever piezoelectric energy harvester. *Smart Mater. Struct.* **2019**, *28*, 085043. [CrossRef]
23. Yang, C.; Chen, K.; Chen, C. Model and Characterization of a Press-Button-Type Piezoelectric Energy Harvester. *IEEE/ASME Trans. Mechatron.* **2019**, *24*, 132–143. [CrossRef]
24. Li, X.; Upadrashta, D.; Yu, K.; Yang, Y. Sandwich piezoelectric energy harvester: Analytical modeling and experimental validation. *Energy Convers. Manag.* **2018**, *176*, 69–85. [CrossRef]
25. Xie, Z.; Wang, T.; Kwuimy, C.K.; Shao, Y.; Huang, W. Design, analysis and experimental study of a T-shaped piezoelectric energy harvester with internal resonance. *Smart Mater. Struct.* **2019**, *28*, 085027. [CrossRef]
26. Li, Z.; Yang, Z.; Naguib, H.; Zu, J. Design and Studies on a Low-Frequency Truss-Based Compressive-Mode Piezoelectric Energy Harvester. *IEEE/ASME Trans. Mechatron.* **2018**, *23*, 2849–2858. [CrossRef]
27. Peng, Y.; Xu, Z.; Wang, M.; Li, Z.; Peng, J.; Luo, J.; Xie, S.; Pu, H.; Yang, Z. Investigation of frequency-up conversion effect on the performance improvement of stack-based piezoelectric generators. *Renew. Energy* **2021**, *172*, 551–563. [CrossRef]
28. Wang, M.; Yin, P.; Li, Z.; Sun, Y.; Ding, J.; Luo, J.; Xie, S.; Peng, Y.; Pu, H. Harnessing energy from spring suspension systems with a compressive-mode high-power-density piezoelectric transducer. *Energy Convers. Manag.* **2020**, *220*, 113050. [CrossRef]
29. Shih, H.-A.; Su, W.-J. Theoretical analysis and experimental study of a nonlinear U-shaped bi-directional piezoelectric energy harvester. *Smart Mater. Struct.* **2018**, *28*, 015017. [CrossRef]
30. Li, Z.; Li, T.; Yang, Z.; Naguib, H.E. Toward a 0.33 W piezoelectric and electromagnetic hybrid energy harvester: Design, experimental studies and self-powered applications. *Appl. Energy* **2019**, *255*, 113805. [CrossRef]
31. Iqbal, M.; Nauman, M.M.; Khan, F.U.; Abas, P.E.; Cheok, Q.; Iqbal, A.; Aissa, B. Multimodal Hybrid Piezoelectric-Electromagnetic Insole Energy Harvester Using PVDF Generators. *Electronics* **2020**, *9*, 635. [CrossRef]
32. Iqbal, M.; Khan, F.U. Hybrid vibration and wind energy harvesting using combined piezoelectric and electromagnetic conversion for bridge health monitoring applications. *Energy Convers. Manag.* **2018**, *172*, 611–618. [CrossRef]
33. Edwards, B.; Aw, K.C.; Hu, A.P.; Tang, L. Hybrid electromagnetic-piezoelectric transduction for a frequency up-converted energy harvester. In Proceedings of the 2015 IEEE International Conference on Advanced Intelligent Mechatronics (AIM), Busan, Korea, 7–11 July 2015; pp. 1149–1154.
34. Pyo, S.; Kwon, D.-S.; Ko, H.-J.; Eun, Y.; Kim, J. Frequency Up-Conversion Hybrid Energy Harvester Combining Piezoelectric and Electromagnetic Transduction Mechanisms. *Int. J. Precis. Eng. Manuf. Green Technol.* **2021**, 1–11. [CrossRef]
35. Li, Z.; Liu, Y.; Yin, P.; Peng, Y.; Luo, J.; Xie, S.; Pu, H. Constituting abrupt magnetic flux density change for power density improvement in electromagnetic energy harvesting. *Int. J. Mech. Sci.* **2021**, *198*, 106363. [CrossRef]
36. Toyabur, R.M.; Salauddin, M.; Cho, H.; Park, J.Y. A multimodal hybrid energy harvester based on piezoelectric-electromagnetic mechanisms for low-frequency ambient vibrations. *Energy Convers. Manag.* **2018**, *168*, 454–466. [CrossRef]
37. Hamid, R.; Yuce, M.R. A wearable energy harvester unit using piezoelectric–electromagnetic hybrid technique. *Sens. Actuators A Phys.* **2017**, *257*, 198–207. [CrossRef]
38. Iqbal, M.; Khan, F.U.; Mehdi, M.; Cheok, Q.; Abas, E.; Nauman, M.M. Power harvesting footwear based on piezo-electromagnetic hybrid generator for sustainable wearable microelectronics. *J. King Saud Univ. Eng. Sci.* **2020**. [CrossRef]
39. Li, Y.; Chen, Z.; Zheng, G.; Zhong, W.; Jiang, L.; Yang, Y.; Jiang, L.; Chen, Y.; Wong, C.-P. A magnetized microneedle-array based flexible triboelectric-electromagnetic hybrid generator for human motion monitoring. *Nano Energy* **2020**, *69*, 104415. [CrossRef]
40. Xia, H.; Chen, R.; Ren, L. Analysis of piezoelectric–electromagnetic hybrid vibration energy harvester under different electrical boundary conditions. *Sens. Actuators A Phys.* **2015**, *234*, 87–98. [CrossRef]
41. Xia, H.; Chen, R.; Ren, L. Parameter tuning of piezoelectric–electromagnetic hybrid vibration energy harvester by magnetic force: Modeling and experiment. *Sens. Actuators A Phys.* **2017**, *257*, 73–83. [CrossRef]
42. Li, P.; Gao, S.; Cai, H.; Wu, L. Theoretical analysis and experimental study for nonlinear hybrid piezoelectric and electromagnetic energy harvester. *Microsyst. Technol.* **2016**, *22*, 727–739. [CrossRef]
43. Halim, M.A.; Kabir, M.H.; Cho, H.; Park, J.Y. A Frequency Up-Converted Hybrid Energy Harvester Using Transverse Impact-Driven Piezoelectric Bimorph for Human-Limb Motion. *Micromachines* **2019**, *10*, 701. [CrossRef] [PubMed]
44. Beyaz, M.İ.; Tat, F.; Özkaya, K.Y.; Özbek, R. Hybrid Magnetic-Piezoelectric Energy Harvester for Power Generation around Waistline During Gait. *J. Electr. Eng. Technol.* **2020**, *15*, 227–233. [CrossRef]
45. Fan, K.; Hao, J.; Tan, Q.; Cai, M. A monostable hybrid energy harvester for capturing energy from low-frequency excitations. *J. Intell. Mater. Syst. Struct.* **2019**, *30*, 2716–2732. [CrossRef]

MDPI

St. Alban-Anlage 66

4052 Basel

Switzerland

Tel. +41 61 683 77 34

Fax +41 61 302 89 18

www.mdpi.com

*Micromachines* Editorial Office

E-mail: micromachines@mdpi.com

www.mdpi.com/journal/micromachines